PUBLICATIONS OF THE ISRAEL ACADEMY

OF SCIENCES AND HUMANITIES

SECTION OF SCIENCES

————

FAUNA PALAESTINA

CRUSTACEA I: AMPHIPODA: HYPERIIDEA

FAUNA PALAESTINA

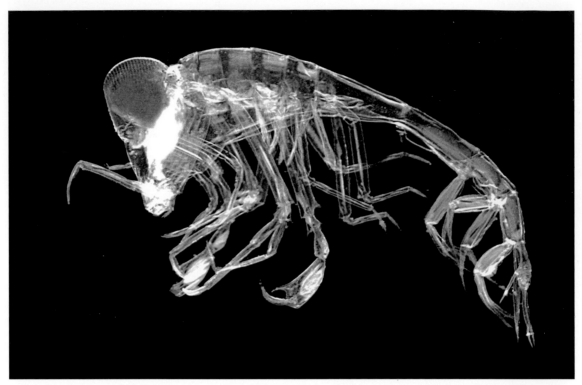

Phronima sedentaria Forskål ♀

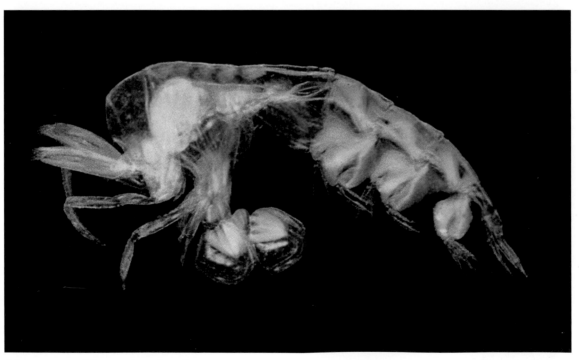

Phronima stebbingi Vosseler ♂

FAUNA PALAESTINA · CRUSTACEA I

AMPHIPODA: HYPERIIDEA OF ISRAEL

A MORPHOLOGICAL ATLAS

by

ENGELINA A. ZELICKMAN

Jerusalem 2005

The Israel Academy of Sciences and Humanities

Authors address:

Dr. Engelina A. Zelickman
3, HaKinor St.
Ma'aleh Adumim
Israel

ISBN 965–208–013–6
ISBN 965–208–164–7

Printed in Israel
at Keterpress Enterprises, Jerusalem

CONTENTS

PLATES

PREFACE

The suborder Hyperiidea (order Amphipoda), contains at least 320 species placed in 72 genera and 23 families. Hyperiidea are marine pelagic predators, ectoparasites or commensals. They feed or live on gelatinous plankton such as Siphonophora, Cnidaria, Ctenophora, Pteropoda, Heteropoda and Thaliacea. They inhabit both cold and tropical seas, down to a depth of 2,700 metres, though most known species live in the upper 500 m. Diel and seasonal vertical migration is common. The deep-sea species are little studied as there are few specimens collected. Hyperiidea form especially dense swarms in the Arctic and Antarctic oceans where they are an important food resource for whales and large fish. For instance, up to 400 kg of freshly caught Hyperiidea were reported from the stomach of one whale. Likewise, up to 2,000 specimens have been found under the umbrella of large cold-water medusae. By feeding on gelatinous plankton, the Hyperiidea are an important factor, linking the often remarkably large biomass of these organisms into the general pelagic food chain.

A total of 105 epipelagic and mesopelagic species have been reported from the Israeli coasts of the Mediterranean and from the northern Red Sea Gulf of Aqaba. These form the subject of the present Atlas. Of the 99 Mediterranean species, 51 are new reports for this sea. Likewise, among the 46 Red Sea species, 37 are reported here for the first time. Of all the 105 species treated here, 40 have been found in both seas.

Hyperiidea morphology and taxonomy are outdated, since high magnification morphological features have never been described before and taken into consideration. In this sense, the present Atlas follows in the footsteps of a previous publication (Zelickman and Por, 1996) which dealt with the micromorphology of the pereiopods in several species of Hyperiidea. Details with possible functional significance and probable taxonomic value are being presented here for the first time. This might be the case particularly with the detailed S.E.M. photographs of the mouthparts.

Hyperiidea are generally 3–4 mm long, with a habitus basically derived from the shrimp-like, caridoid shape. However several species of *Phronima* and *Parathemisto* may reach 40–60 mm in length. The epipelagic, surface species, are transparent bluish-white; the deep-sea species are predominantly red or brown in colour.

The head, or cephalon, may have an inflated globose, conical or acicular shape. Characteristically, the eyes may occupy the whole surface of the cephalon. The eyes are somewhat reduced in some bathyal and bathypelagic species.

The head bears six sets of paired appendages: the antennulae (A1) in the front, followed by the antennae (A2), the mandibulae (Md), the maxillulae (Mx1), the maxillae (Mx2) and the maxillipeds (Mxp). The four last pairs of cephalic appendages are joined into an oral cone, and serve as mouthparts. The antennulae and the antennae may be partially reduced or even lost, especially in the adult females. The antennalae usually have a three-jointed peduncle, bearing a main flagellum. The accessory flagellum, present in other Amphipoda, is absent in the Hyperiidea. The segments of the peduncule may be fused. The antennae have a flagellum of cylindrical joints, presenting complicated cuticular zig-zag structures. The mouthparts of the Hyperiidea are greatly reduced in comparison to the other Amphipoda and the type and degree of reduction varies from family to family.

The thorax (pereion) consists of seven segments, each with a pair of walking legs, the pereiopods. The coxal plates at the base of the pereiopods are small. The abdomen (pleon) consists of six segments. The first three segments bear biramous, segmented swimming legs, the pleopods. Each of the last three abdominal segments, also called the urosoma, bear a pair of biramous, unsegmented uropods. The abdomen ends in a plate, the telson.

The most important features used in Hyperiidea taxonomy are the shape of the head and the appendages, and especially the detailed morphology of the first two pereiopods. Other features of importance are the length/width ratio and the spinulation of the pereiopod joints. Several genera have their first pereiopod transformed into a powerful prehensile subchela, resulting from the fusion of the sixth and seventh leg joints. In the Platyscelidae, the second joints of pereiopods 5 and 6 are much broadened. As a result, they cover the bent abdomen and impart a ball-like appearance to the whole animal. In this situation, the distal joints of pereiopods 6 and 7 may be reduced. As a rule, all the uropods are biramous, but unlike other Amphipoda, the exo- and endopodite branches are never biarticulated. In some cases the basipodite and one of the branches may be reduced.

The Hyperiidea have separate sexes. Reproduction is seasonal in most of the species, though some species reproduce around the year. As a rule, the female carries several tens of eggs and embryos between its coxal plates. Development is direct, without metamorphosis. Sexual dimorphism and age-related meristic and morphological changes are often a source of error in taxonomic treatments.

While no regional illustrated monograph of the Hyperiidea of the Mediterranean and the Red Seas exists, most of the species are cosmopolitan and there is considerable geographic variability. Most of the literature sources are expedition reports in which illustration is schematic and of poor quality. In some cases, illustration details have been combined and borrowed from various earlier publications.

In the present Atlas I have tried to present, as consistently as possible, original drawings of the species resulting from the study of a single specimen, or more rarely, from several specimens from the same sample. Furthermore, in addition to the classical characters used in Hyperiidea taxonomy, I have added important structural details of the mouthparts and legs.

Anatomical notes accompany the drawings on the facing pages. The Bibliography at the end of the Atlas results from a thorough search of all the publications on the taxonomy of the Hyperiidea. I believe that the present Atlas can serve as a basic tool for identification of the Hyperiidea the world over.

ACKNOWLEDGEMENTS

I am very grateful to the Department of Evolution, Systematics and Ecology, the Hebrew University of Jerusalem, for the opportunity to conduct my research. Special thanks are due to Prof. Francis Dov Por for sponsoring my research and putting the Aquatic Invertebrates Collections of that University at my disposal. I am grateful to the late Prof. Zeev Reiss for his DCPE collections from the Red Sea and to Prof. Baruch Kimor of the Technion in Haifa for the Mediterranean plankton samples. Many thanks are due to Dr. Heather Bromley and Ms Zofia Lasman for the careful editorial work. I am grateful to Dr. Chanan Dimentman who kindly and competently checked the bibliography. The cover illustration, the frontispiece and the S.E.M. photographic plates are a testimony to the photographic skills of David Darom.

Financial support for my research was provided by the Ministry of Absorption and by the Israel Academy of Sciences and Humanities through the "National Collections of Natural History", and the "Fauna et Flora Palaestina Committee".

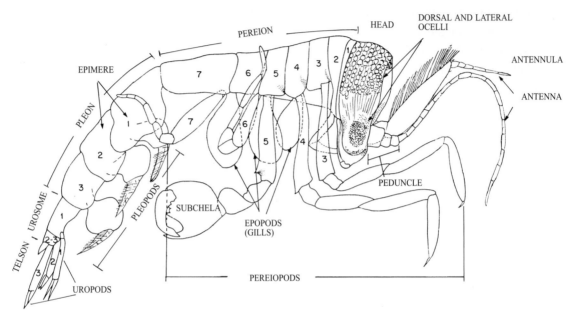

Anatomical Schema of a Hyperiidea
(Bowman and Gruner, 1973)

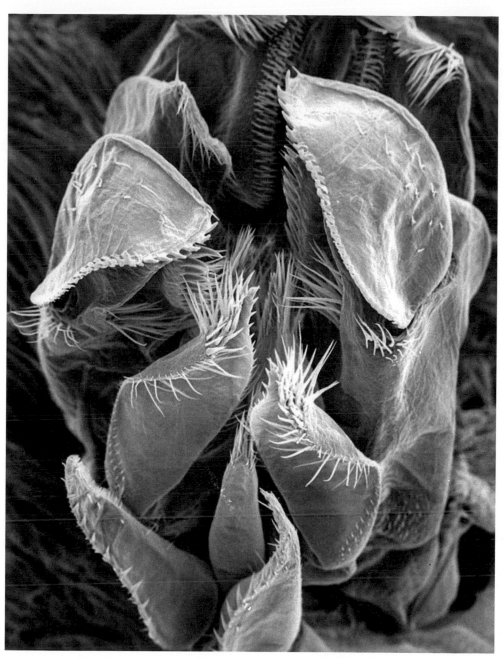

General view of mouthparts complex
Phronima sedentaria Forskål, 1775
(See Plate I:A)

UPDATED LIST OF HYPERIIDEA SPECIES
KNOWN FROM THE MEDITERRANEAN AND RED SEAS

Taxonomic order in this list is after M.E. Vinogradov et al, 1982.

Key to symbols

M   species previously reported in the Mediterranean Sea
M*   my findings in the Mediterranean Sea
R   species previously reported in the Red Sea
R*   my findings in the Red Sea

Infraorder PHYSOSOMATA Pirlot, 1929
  Super family SCINOIDEA Stebbing, 1888
    Family SCINIDAE Stebbing, 1888

1.   *Scina crassicornis* (Fabricus, 1775). M*, R*
2.   *S. curvidactyla* Chevreux, 1914. M, R
3.   *S. borealis* (G.O. Sars, 1880). M*
4.   *S. spinosa* Vosseler, 1901. M*
5.   *S. stebbingi* Chevreux, 1919. M
6.   *S. marginata* (Bovallius, 1885). M*
7.   *S. rattrayii rattrayii* Stebbing, 1895. M
8.   *S. stenopus* Stebbing, 1895. M
9.   *S. tullbergi* (Bovallius, 1885). M, R*
10. *S. similis* Stebbing, 1895. M
11. *S. alberti* Chevreux, 1919. M

Infraorder PHYSOCEPHALATA Bowman et Gruner, 1973
  Super family VIBILIOIDEA Dana, 1852
    Family VIBILIIDAE Dana, 1852

12. *Vibilia jeangerardi* (Lucas, 1895). M, R*
13. *V. viatrix* Bovallius, 1887. M*, R*
14. *V. australis* Stebbing, 1888. M
15. *V. stebbingi* Behning et Woltereck, 1912. M*
16. *V. gibbosa* Bovallius, 1887. M*
17. *V. borealis* Bate et Westwood, 1868. M
18. *V. chuni* Behning, 1912. R
19. *V. armata* Bovallius, 1887. M*
20. *V. cultripes* Vosseler, 1901. M

Family PARAPHRONIMIDAE Bovallius, 1887
21. *Paraphronima gracilis* Claus, 1879. M, R*
22. *P. crassipes* Claus, 1879. M, R*

Superfamily PHRONIMOIDEA Dana, 1872
Family HYPERIIDAE Dana, 1872
23. *Iulopis loveni* Bovallius, 1887. M, R
24. *Hyperoche mediterranea* Senna, 1908. M
25. *H. picta* Bovallius, 1889. M
26. *Parathemisto gaudichaudi* (Guérin-Ménéville, 1825). M
27. *Bougisia ornata* Laval, 1966. M
28. *Hyperioides longipes* (Chevreux, 1900). M*, R*
29. *H. sibaginis* (Stebbing, 1888). M*
30. *Lestrigonus schizogeneios* (Stebbing, 1888). M*, R*
31. *L. shoemakeri* Bowman, 1973. M*
32. *L. crucipes* (Bovallius, 1889). M*
33. *L. macrophthalmus* (Vosseler, 1901). M*
34. *L. latissimus* (Bovallius, 1889). M* ,R*
35. *L. bengalensis* Giles, 1887. M*
36. *Hyperietta luzoni* (Stebbing, 1888). M, R*
37. *H. vosseleri* (Stebbing, 1904). M*
38. *H. stephenseni* Bowman, 1973. M
39. *Hyperionyx macrodactylus* (Stephensen, 1924). M*
40. *Phronimopsis spinifera* Claus, 1879. M*, R*

Family DAIRELLIDAE Bovallius, 1887
41. *Dairella latissima* Bovallius, 1887. M

Family PHRONIMIDAE Dana, 1852
42. *Phronima sedentaria* Forskål, 1775. M*, R*
43. *Ph. atlantica* Guérin-Ménéville, 1836. M, R*
44. *Ph. solitaria* Guérin-Ménéville, 1836. M ,R
45. *Ph. stebbingi* Vosseler, 1901. M*
46. *Ph. curvipes* Vosseler, 1901. M*
47. *Ph. colletti* Bovallius, 1887. M*
48. *Ph. pacifica* Streets, 1877. M, R*
49. *Phronimella elongata* (Claus, 1862). M*

Family PHROSINIDAE Dana, 1853
50. *Phrosina semilunata* Risso, 1822. M ,R*
51. *Anchylomera blossevillei* Milne Edwards, 1830. M, R*
52. *Primno macropa* Guérin-Ménéville, 1836. M, R*
53. *P. brevidens* Bowman, 1978. R*
54. *P. lattreillei* Stebbing, 1888. M, R*

Superfamily LYCAEOPSOIDEA Chevreux, 1913
    Family LYCAEOPSIDAE Chevreux, 1913
55. *Lycaeopsis themistoides* Claus, 1897. M*, R*
56. *L. zamboangae* (Stebbing, 1888). R

Superfamily PLATYSCELOIDEA Bate, 1862
    Family PRONOIDAE Claus, 1879
57. *Eupronöe maculata* Claus, 1879. M, R*
58. *E. minuta* Claus, 1879. M*
59. *E. armata* Claus, 1879. R*
60. *Pronöe capito* Guérin-Ménéville, 1836. M
61. *Parapronöe parva* Claus, 1879. M*
62. *Paralycaea gracilis* Claus, 1879. M*

    Family ANAPRONOIDAE Bowman et Gruner, 1973
63. *Anapronöe reinhardti* Stephensen, 1925. M

    Family LYCAEIDAE Claus, 1879
64. *Lycaea pulex* Marion, 1874. M*
65. *L. pauli* Stebbing, 1888. M*
66. *L. serrata* Claus, 1879. M*
67. *L. pachypoda* (Claus, 1879). M
68. *Simorhynchotus antennarius* (Claus, 1871). M, R*

    Family TRYPHANIDAE Bovallius, 1887
69. *Tryphana malmi* Boeck, 1870. M

    Family BRACHYSCELIDAE Stephensen, 1923
70. *Brachyscelus crusculum* Bate, 1861. M, R*
71. *B. globiceps* (Claus, 1879). M*
72. *B. rapax* Claus, 1879. M*
73. *B. macrocephalus* Stephensen, 1925. M*
74. *Euthamneus rostratus* (Bovallius, 1887). M*

    Family OXYCEPHALIDAE Bate, 1861
75. *Oxycephalus piscator* Milne Edwards, 1830. M*
76. *O. clausi* Bovallius, 1887. M*
77. *O. latirostris* Claus, 1879. M, R
78. *O. longipes* Spandl, 1927. M
79. *Streetsia challengeri* Stebbing, 1888. M*, R*
80. *S. steenstrupi* (Bovallius, 1887). M, R*
81. *S. porcella* (Claus, 1879). M*
82. *S. mindanaonis* (Stebbing, 1888). M
83. *Leptocotis tenuirostris* (Claus, 1871). M*
84. *Calamorhynchus pellucidus* Streets, 1878. M*
85. *Glossocephalus milne edwardsi* Bovallius, 1887. M*, R*

86. *Cranocephalus scleroticus* (Streets, 1878). M*
87. *Rhabdosoma whitei* Bate, 1862. M*, R*
88. *R. brevicaudatum* Stebbing, 1888. M*

Family PLATYSCELIDAE Bate, 1862
89. *Platyscelus ovoides* (Risso, 1816). M*, R*
90. *P. armatus* (Claus, 1879). M, R*
91. *P. serratulus* Stebbing, 1888. M*, R*
92. *Hemityphis rapax* (Milne-Edwards, 1836). M
93. *Paratyphis maculatus* Claus, 1879. M*
94. *P. spinosus* Spandl, 1924. M*, R*
95. *P. promontorii* Stebbing, 1888. R*
96. *Tetrathyrus forcipatus* Claus, 1879. M, R*
97. *T. arafurae* Stebbing 1888. M*
98. *Amphithyrus bispinosus* Claus, 1879. M
99. *A. similis* Claus, 1879. M, R
100. *A. glaber* Spandl, 1924. R
101. *A. sculpturatus* Claus, 1879. M*, R*

Family PARASCELIDAE Claus, 1879
102. *Schizoscelus ornatus* Claus, 1879. M
103. *Thyropus sphaeroma* (Claus, 1879). M, R
104. *Parascelus typhoides* Claus, 1879. M*, R*
105. *P. edwardsi* Claus, 1879. M, R*

MORPHOLOGICAL ATLAS

FIGURE 1a

SCINIDAE Stebbing

Scina crassicornis (Fabricius, 1775) ("crassicornis group")

Female, 18 mm

1 Lateral view
2 Maxillula
3 Maxilla
4 Maxilliped
5 Maxilliped, tip of inner lobe
6 Maxilliped, tip of outer lobe

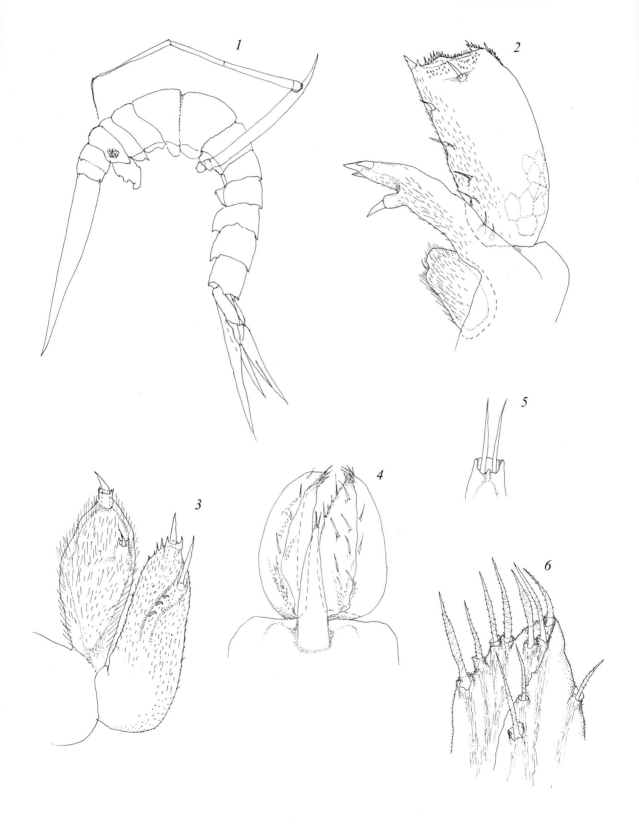

FIGURE 1b

Scina crassicornis (Fabricius, 1775) (cont.)

7 Pleopod 1, bifid spine
8 Pleopod 1, tip of bifid spine
9 Coupling spines of pleopods

Female, 9 mm

10 Mandibula
11 Maxillula

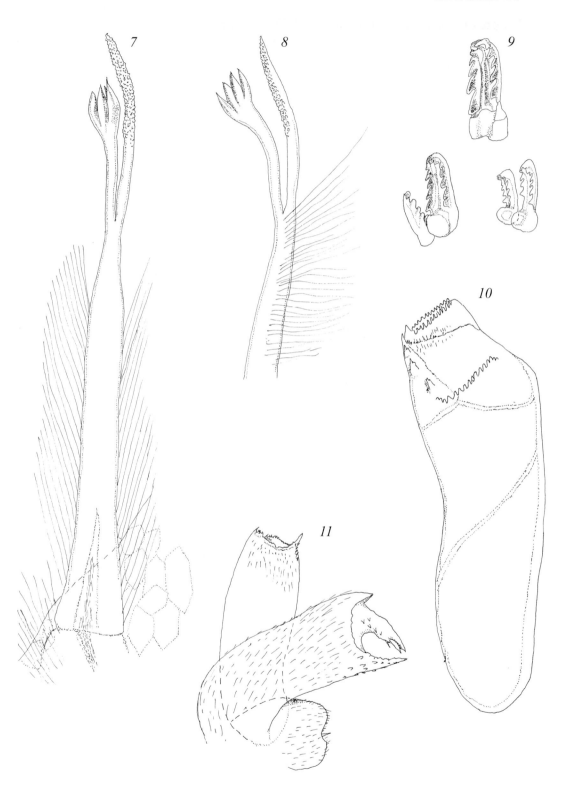

FIGURE 1c

Scina crassicornis (Fabricius, 1775) (cont.)

12 Maxilla, with details
13 Maxilliped, with detail
14 Pereiopod 1, dactyl
15 Pereiopod 7, tip of dactyl
16 Coupling hooks on pleopod 2

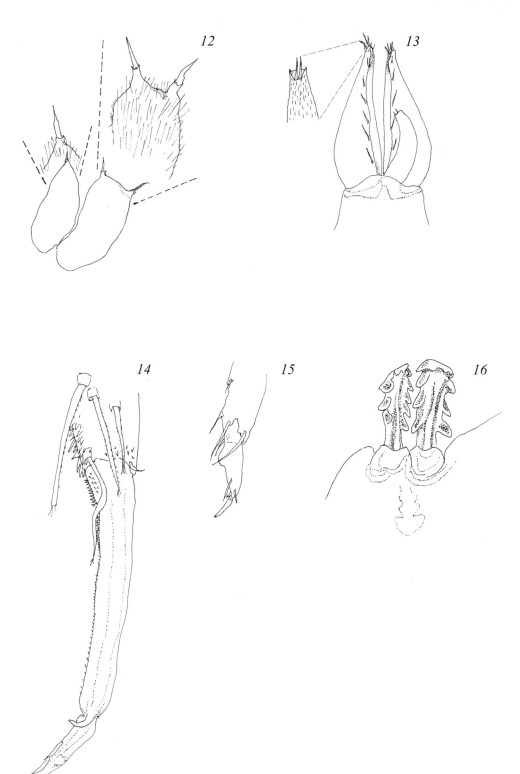

FIGURE 2a

Scina spinosa Vosseler, 1901 ("spinosa group")

Female, 9 mm; males, 7 and 9 mm

1   Female, dorsal view
2   Male, dorsal view
3   Male, lateral view
4   Antennula, female
5   Antennula, male
6   Antenna, male

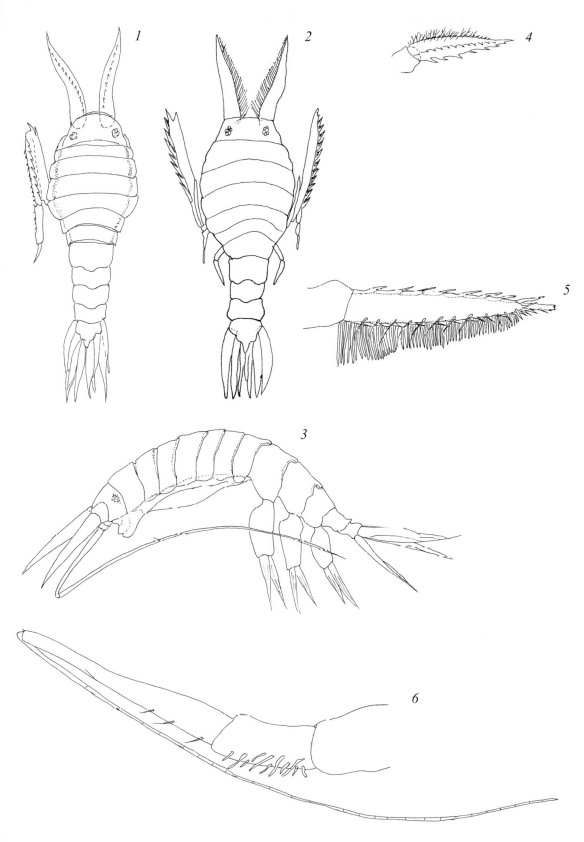

FIGURE 2b

Scina spinosa Vosseler, 1901 (cont.)

7   Mandibula, male
8   Maxilliped, female (flattened)
9   Pereiopod 1, male
10 Pereiopod 2, male
11 Pereiopod 5, female
12 Pereiopod 5, male, with detail
13 Uropods and telson, with details

FIGURE 2b

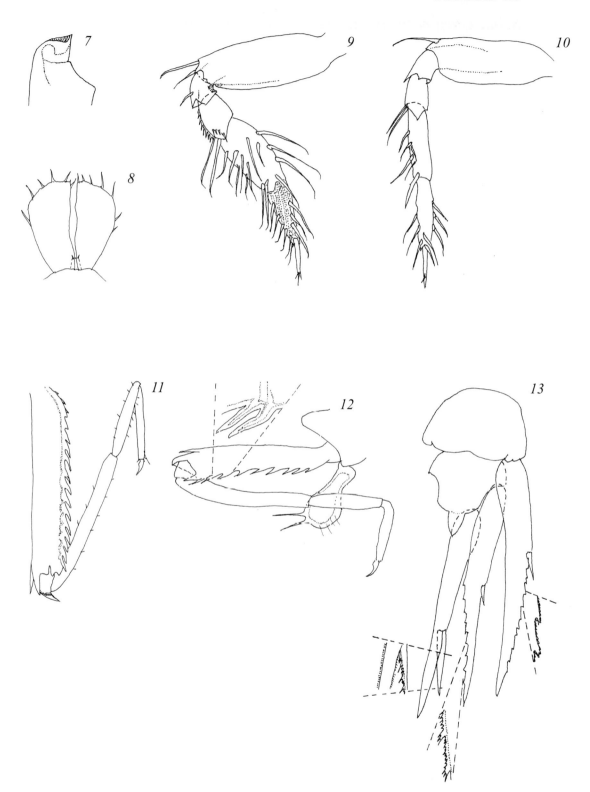

FIGURE 3

Scina marginata (Bovallius, 1885) ("marginata group")

Male, 10 mm

1   Mandibula
2   Maxillula, with detail
3   Maxilla
4   Maxilliped, with detail
5   Pereiopod 1
6   Pereiopod 7
7   Pleopod 2, tip of bifid spine
8   Coupling hooks of pleopod 2
9   Coupling hooks of pleopod 3, with detail

FIGURE 3

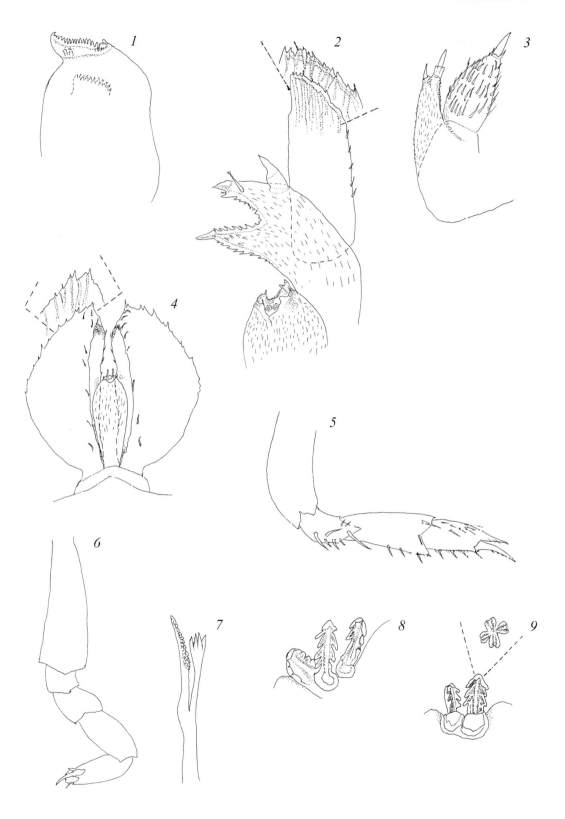

FIGURE 4a

VIBILIIDAE Dana

Vibilia jeangerardi (Lucas, 1895)

Male (?) moulting specimen, 6 mm

1   Lateral view
2   Antennula
3   Antenna
4   Mandibula
5   Maxillula
6   Maxilla

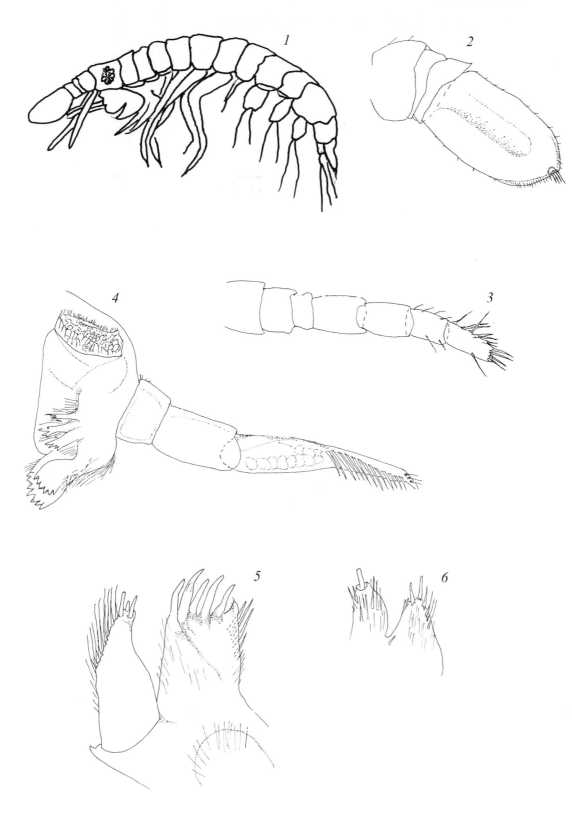

FIGURE 4b

Vibilia jeangerardi (Lucas, 1895) (cont.)

7   Maxilliped
8   Maxilliped, detail of inner lobe
9   Pereiopod 1, with details
10 Pereiopod 2, with details
11 Pereiopod 3
12 Pereiopod 5, with detail

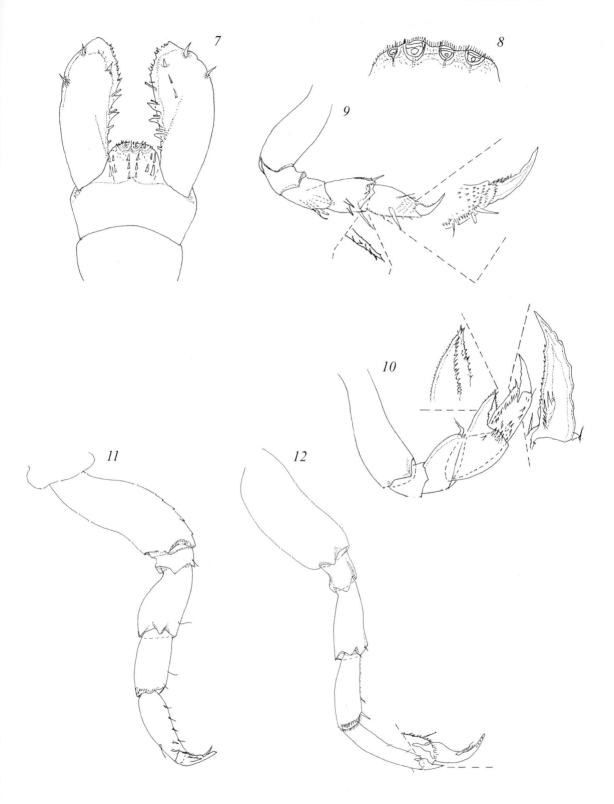

FIGURE 4c

Vibilia jeangerardi (Lucas, 1895) (cont.)

13 Pereiopod 6, with detail
14 Pereiopod 7, with details
15 Pleopod 1, bifid seta
16 Coupling hooks of Pleopod 2
17 Uropods and telson, with details

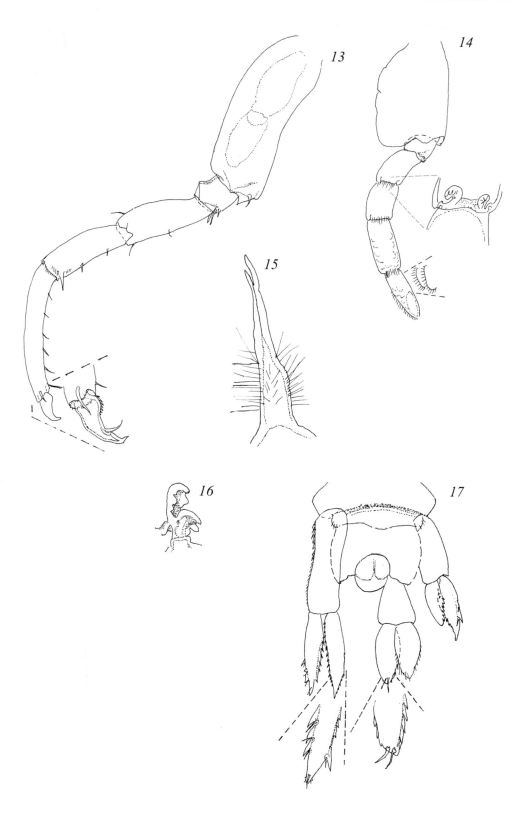

FIGURE 5a

Vibilia viatrix Bovallius, 1887

Females, 8–9 mm

1 Lateral view
2 Antennula, inner face, with details
3 Antennula, outer face, with details
4 Antenna
5 Mandibula, with mandibular palp
6 Mandibula, gnathobase in detail, with molar plate
7 Mandibula, lacinia mobilis, with details

FIGURE 5a

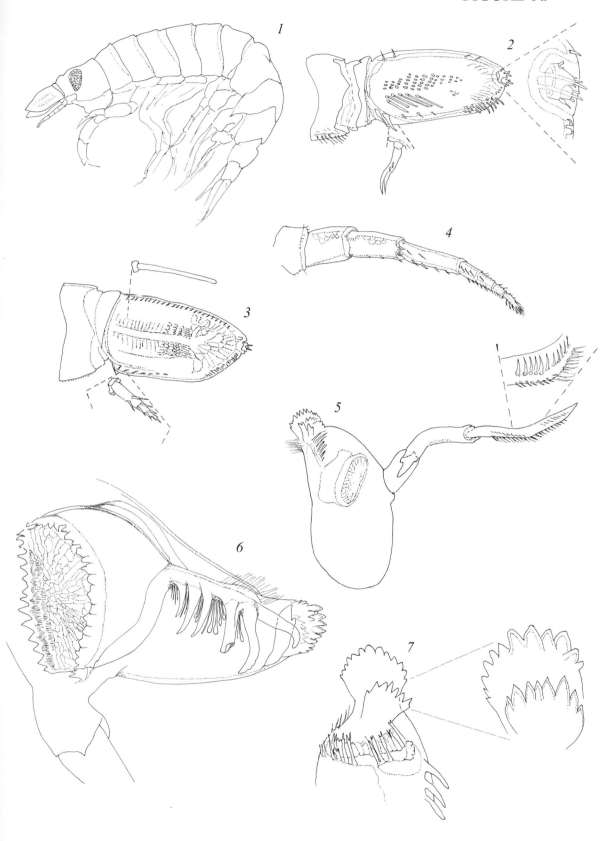

FIGURE 5b

Vibilia viatrix Bovallius, 1887 (cont.)

8   Mandibula, "inferolateralia"
    (*sive* Coleman, 1994, fig. 2AB, p. 349; fig. 9, p. 356)
9   Right maxillula
10 Maxillula, tip of palp
11 Right maxilla
12 Left maxilla, with detail
13 Maxilliped, ventral face, with details
14 Maxilliped, dorsal face, with details

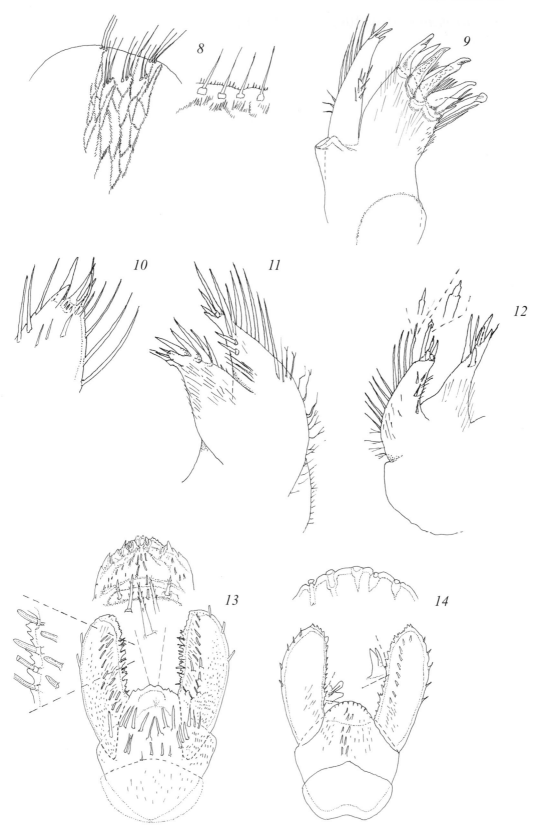

FIGURE 5c

Vibilia viatrix Bovallius, 1887 (cont.)

15 Pereiopod 1, with details
16 Pereiopod 1, another view, with details
17 Pereiopod 2, with details

FIGURE 5c

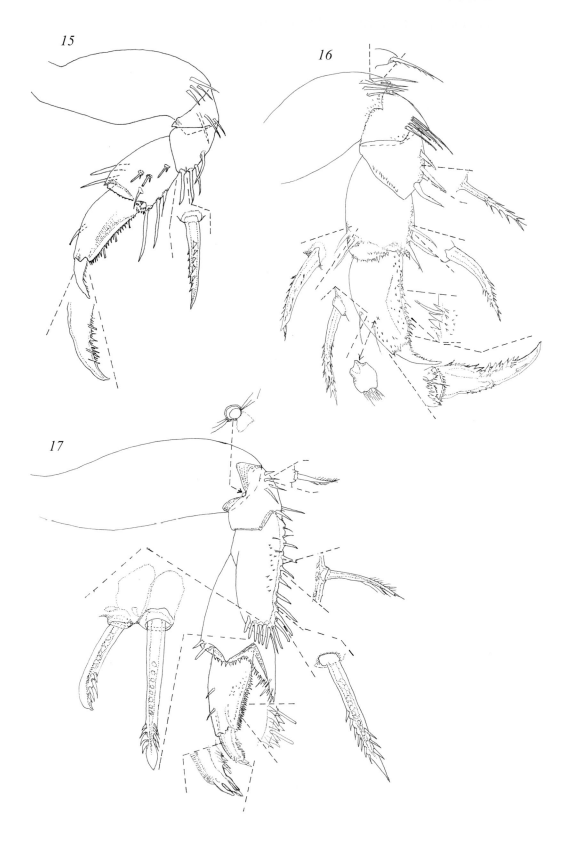

FIGURE 5d

Vibilia viatrix Bovallius, 1887 (cont.)

18 Pereiopod 3
19 Pereiopod 4
20 Pereiopod 5, with details

FIGURE 5e

Vibilia viatrix Bovallius, 1887 (cont.)

21 Pereiopod 6, with details
22 Pereiopod 7, with details
23 Pereiopod 7, with details
24 Pleopod 1, with detail of bifid spine

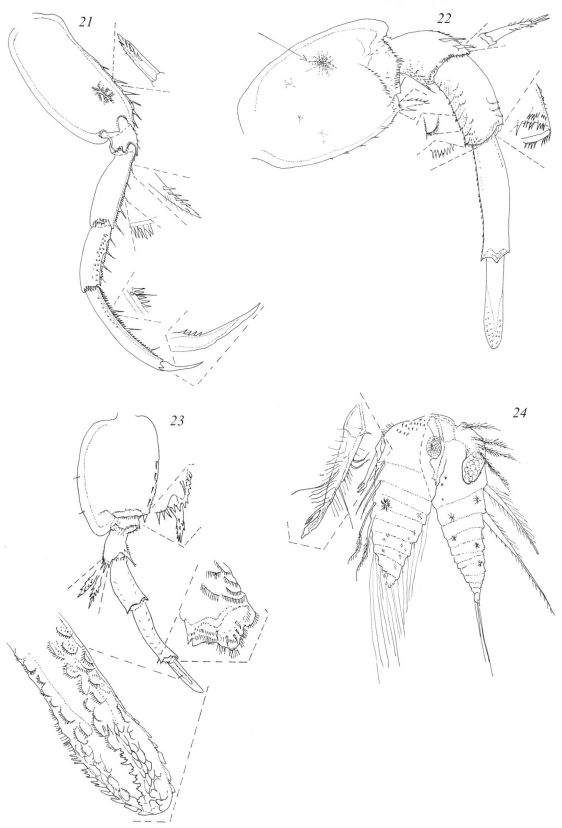

FIGURE 5f

Vibilia viatrix Bovallius, 1887 (cont.)

25 Exopod of pleopod 2, with detail of bifid spine
26 Coupling hooks of pleopods 1–3
27 Uropods and telson, with detail

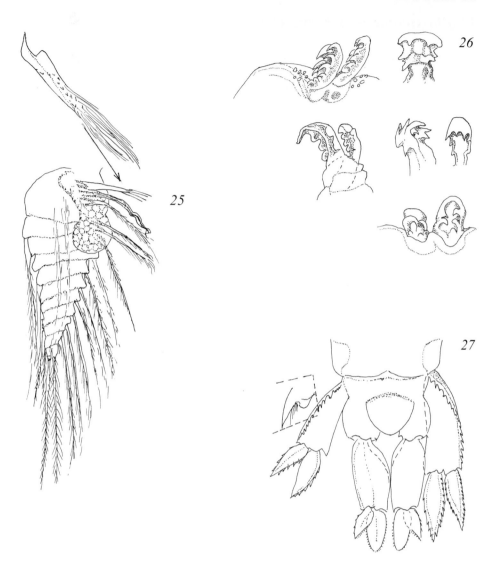

25

26

27

FIGURE 6a

Vibilia stebbingi Behning et Woltereck, 1912

Females, 5–6 mm; males, 5–6 mm

1   Female, lateral view
2   Antennula, male
3   Antennula, male, another view
4   Antenna, male
5   Mandibula and palp
6   Maxillula
7   Maxilla

FIGURE 6a

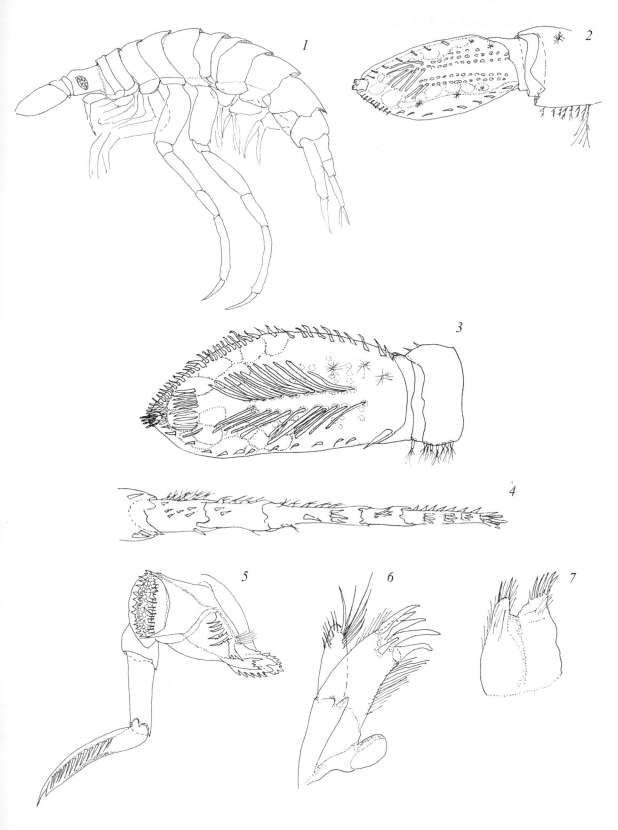

[33]

FIGURE 6b

Vibilia stebbingi Behning et Woltereck, 1912 (cont.)

8   Maxilliped, male
9   Maxilliped, female
10 Pereiopod 1, female, with detail
11 Pereiopod 1, male
12 Pereiopod 2, male; the dactyl belongs to another specimen
13 Pereiopod 3, male
14 Pereiopod 4, male

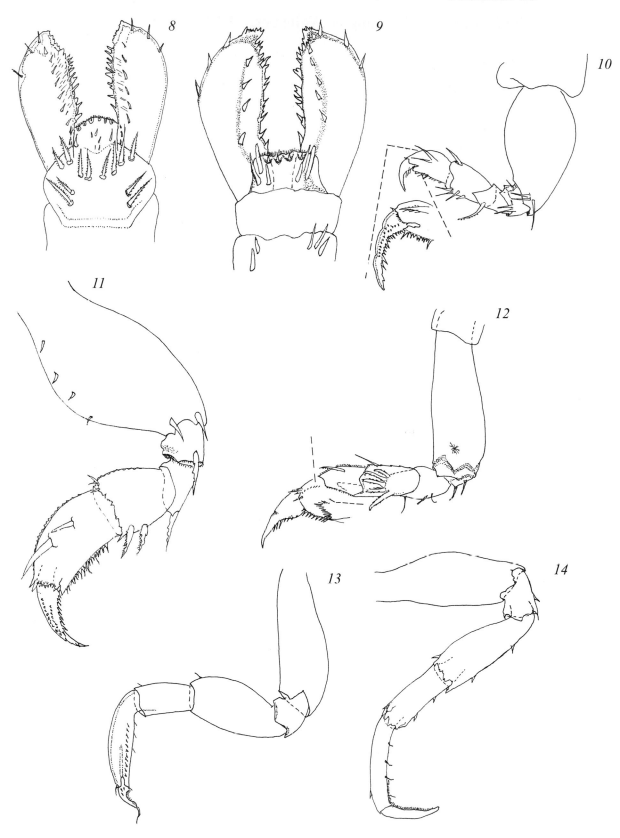

FIGURE 6c

Vibilia stebbingi Behning et Woltereck, 1912 (cont.)

15 Pereiopod 5, male
16 Pereiopod 6, male, with detail
17 Pereiopod 7, male, with detail
18 Pereiopod 7, male, another view, with detail
19 Pereiopod 7, female, with detail

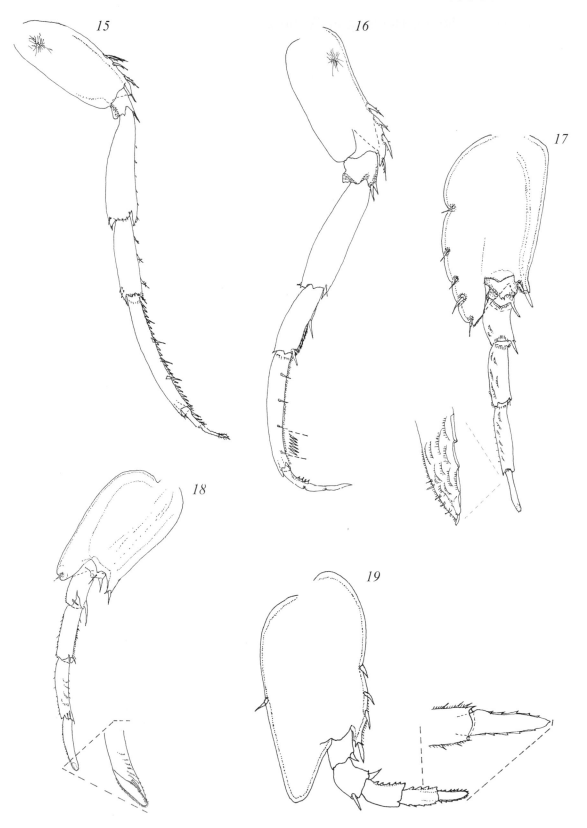

FIGURE 6d

Vibilia stebbingi Behning et Woltereck, 1912 (cont.)

20 Coupling hooks of pleopods 1 and 2
21 Uropods and telson, female
22 Uropods and telson, male, with details

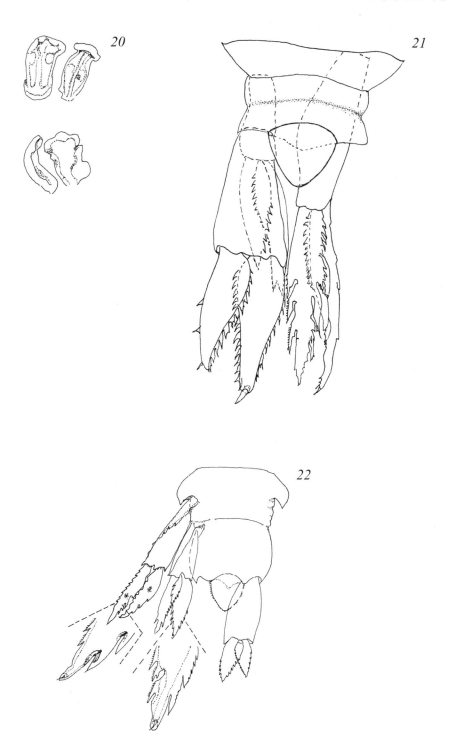

FIGURE 7a

Vibilia gibbosa Bovallius, 1887

Female (immature?), 4 mm

1 Lateral view
2 Antennula, external view
3 Antenna
4 Mandibula with palp and detail
5 Maxillula
6 Maxilla
7 Maxilliped, frontal view
8 Pereiopod 1
9 Pereiopod 2, with detail

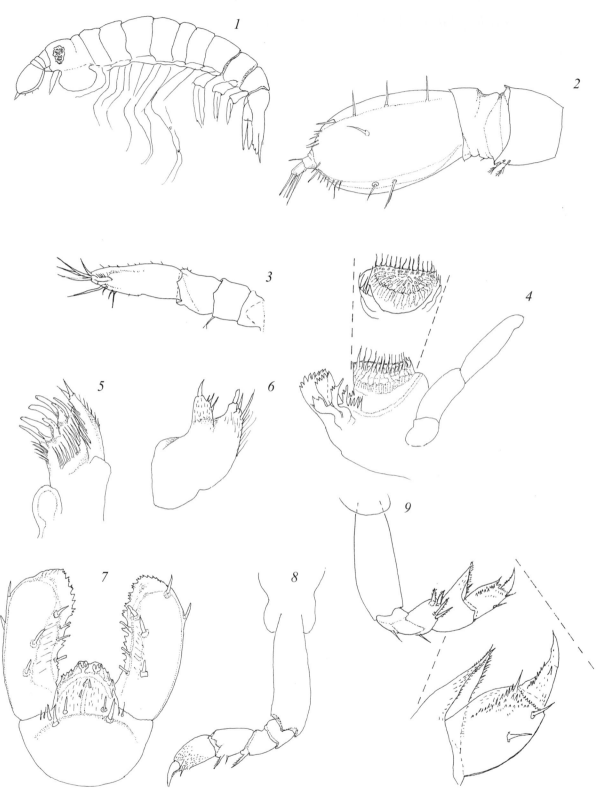

FIGURE 7b

Vibilia gibbosa Bovallius, 1887 (cont.)

10 Pereiopod 4, with detail
11 Pereiopod 5
12 Pereiopod 6
13 Pereiopod 7, with detail
14 Coupling hooks and bifid seta
15 Uropods and telson

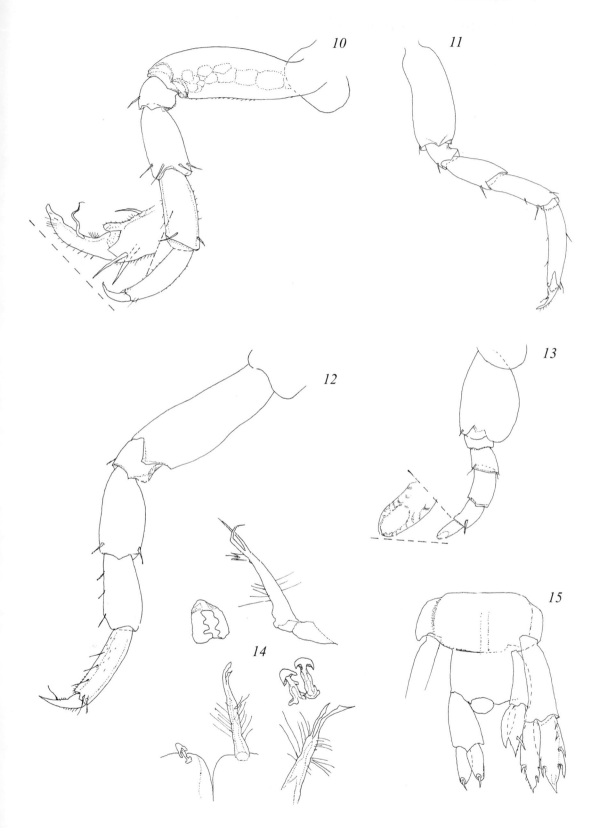

10

11

12

13

14

15

FIGURE 8a

Vibilia armata Bovallius, 1887

Female, 8 mm

1   Lateral view
2   Antennula, ventro-lateral view
3   Antennula, external side, with detail
4   Antenna, with detail
5   Mandibular complex, with details
6   Molar plate of mandibula
7   Inferolateralia, with details

FIGURE 8a

1

2

3

4

5

6

7

FIGURE 8b

Vibilia armata Bovallius, 1887 (cont.)

8a Maxillula, anterior view
8b Maxillula, posterior view
9 Maxilla, with detail
10 Maxilliped (spinulation partly omitted), with detail
11 Maxilliped, median lobe with styliform spines partly omitted
12 Pereiopod 1, with detail
13 Pereiopod 2, with details

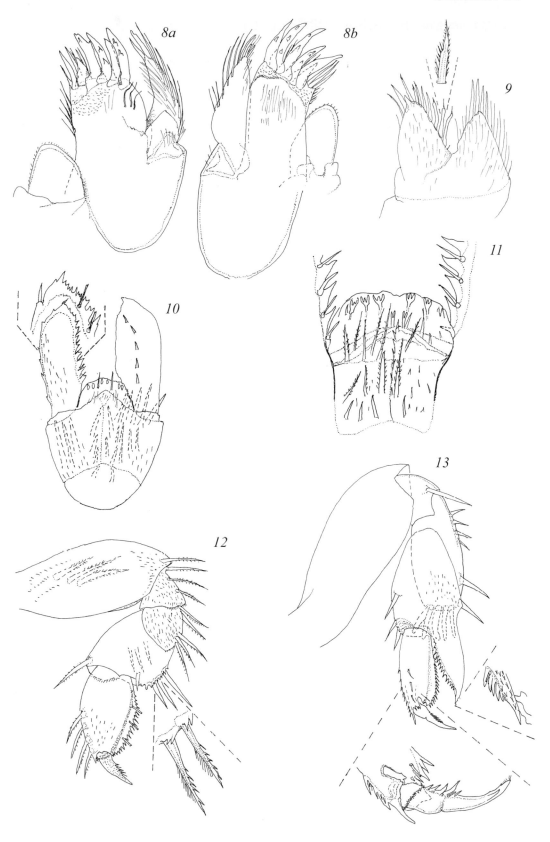

FIGURE 8c

Vibilia armata Bovallius, 1887 (cont.)

14 Pereiopod 5, with details
15 Pereiopod 6, with details
16 Pereiopod 7
17 Last segments of pereiopod 7, with details

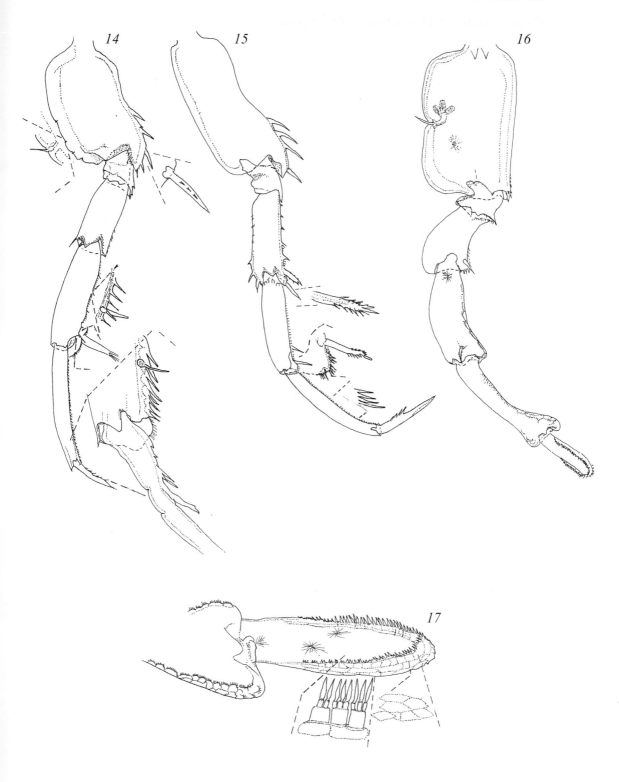

FIGURE 8d

Vibilia armata Bovallius, 1887 (cont.)

18 Telson and uropods, with details
19 Bifid spine and coupling hooks on pleopod 1

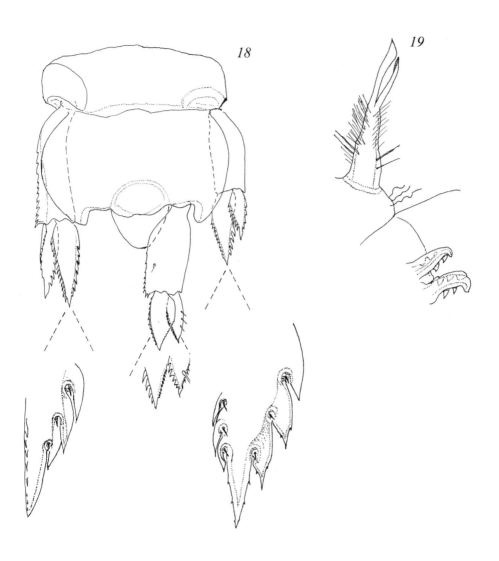

18

19

FIGURE 9a

PARAPHRONIMIDAE Bovallius

Paraphronima gracilis Claus, 1879

Female, 21 mm

1   Lateral view
2   Antennula
3   Antenna
4   Mandibula
5   Maxillula
6   Maxilliped

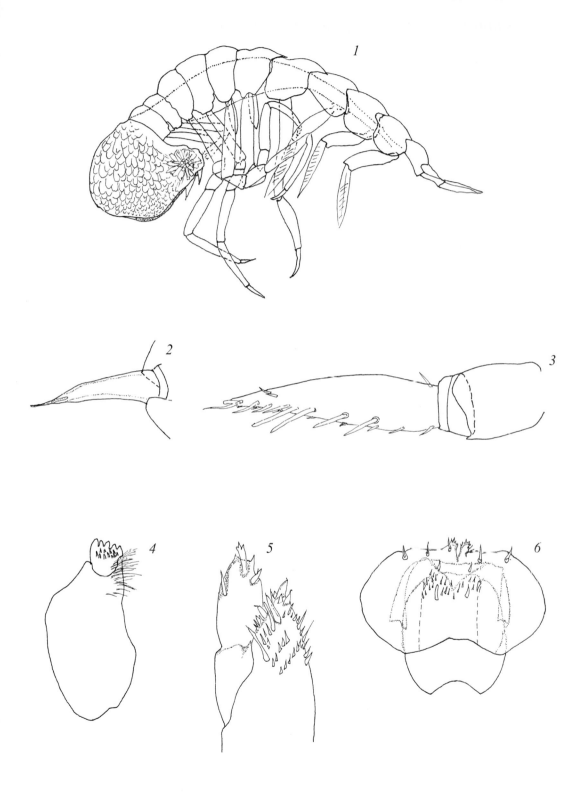

FIGURE 9b

Paraphronima gracilis Claus, 1879 (cont.)

7 Pereiopod 1, with detail of dactylus
8 Pereiopod 2, with detail of dactylus
9 Pereiopod 3, with detail of dactylus
10 Pereiopod 4, with detail of dactylus

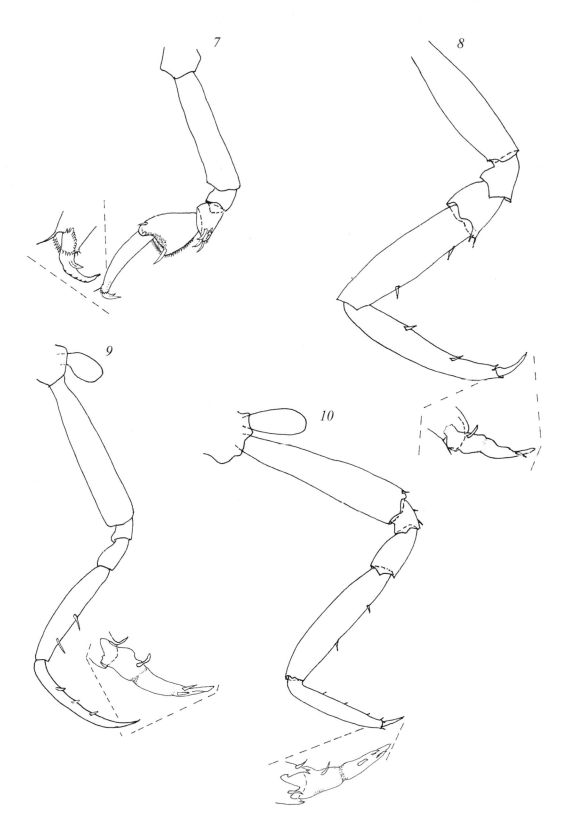

FIGURE 9c

Paraphronima gracilis Claus, 1879 (cont.)

11 Pereiopod 5, with detail of dactylus
12 Pereiopod 6
13 Pereiopod 7, with detail of dactylus
14 Uropods and telson

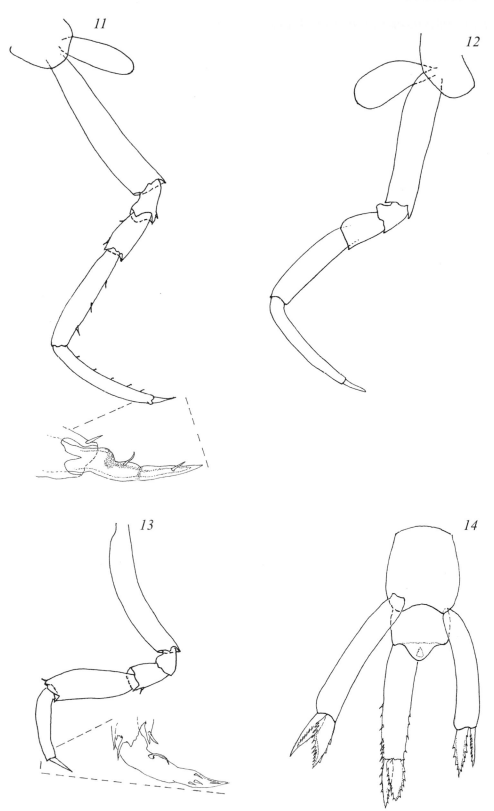

FIGURE 10a

Paraphronima crassipes Claus, 1879

Males, 8 and 12 mm

1   Lateral view
2   Antennula, internal surface
3   Antenna, with details
4   Mandibula
5   Maxillula, male 8 mm
6   Maxillula, frontal view, male 12 mm
7   Maxilla, male 8 mm
8   Maxilliped, lateral view

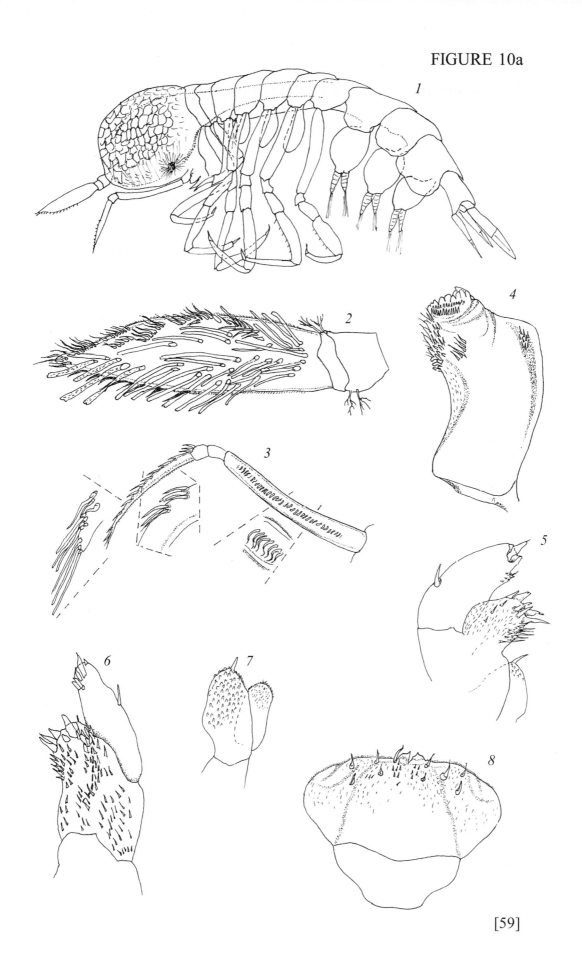

FIGURE 10b

Paraphronima crassipes Claus, 1879 (cont.)

9   Pereiopod 1, with detail
10 Pereiopod 2, with two detailed views of the dactylus (face and profile)
11 Pereiopod 3, with detail of dactylus
12 Pereiopod 4, with detail of dactylus

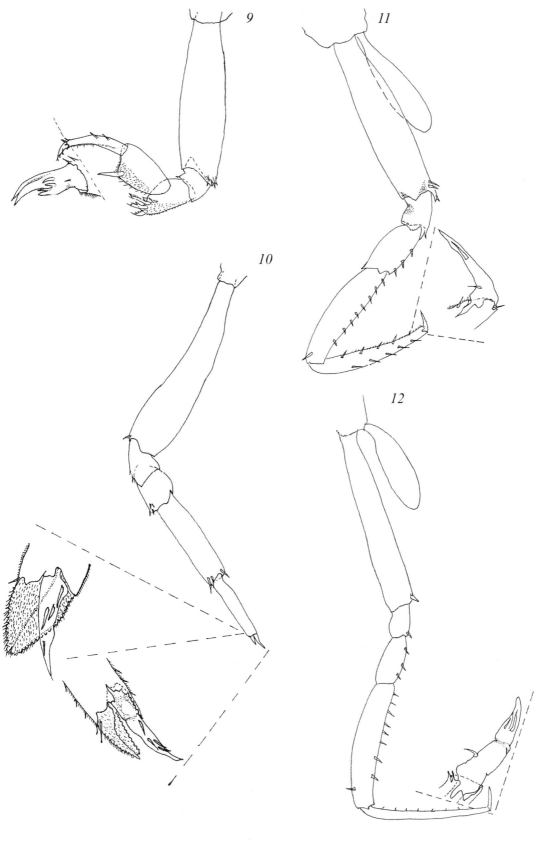

FIGURE 10c

Paraphronima crassipes Claus, 1879 (cont.)

13 Pereiopod 5, with detail of dactylus
14 Pereiopod 6, with detail of dactylus
15 Pereiopod 7, with detail of dactylus
16 Uropods and telson, with detail

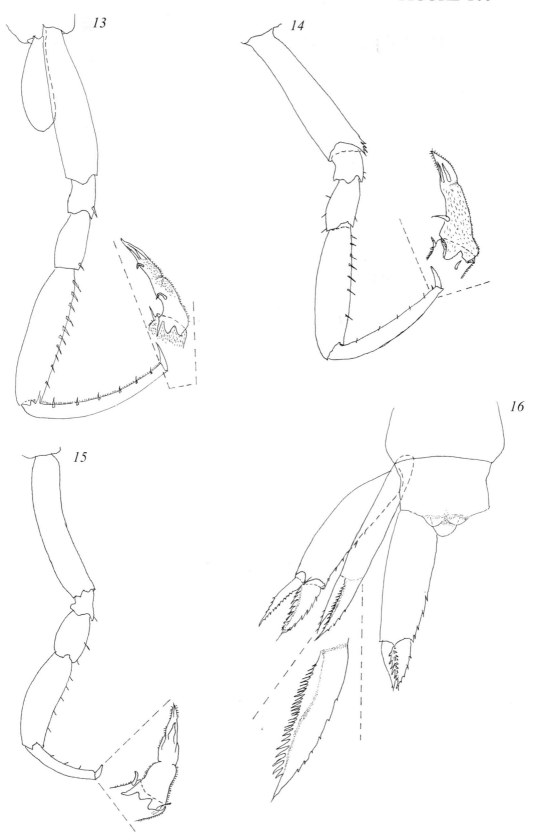

FIGURE 11a

HYPERIIDAE Dana

Hyperioides longipes (Chevreux, 1900)

Female, 4 mm

1   Detail of head, lateral view
2   Antennula
3   Antenna
4   Mandibula
5   Maxillula, right
6   Maxillula, left
7   Maxilla
8   Maxilliped, anterior

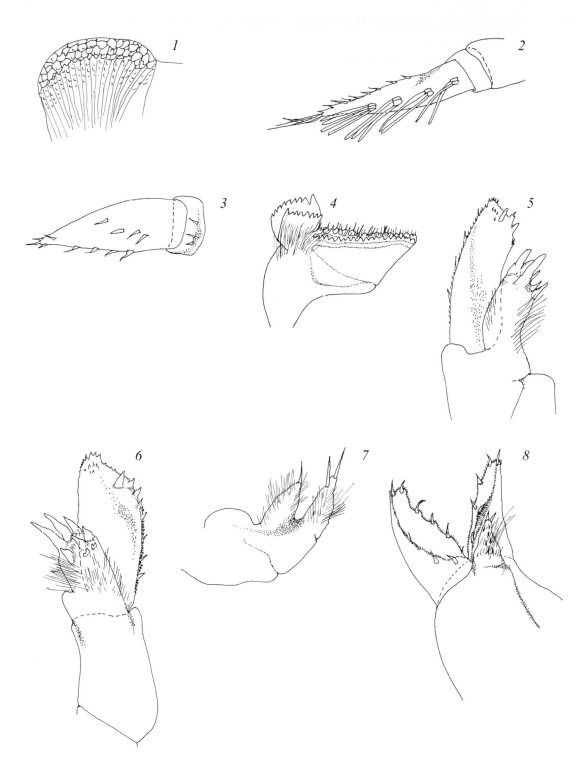

FIGURE 11b

Hyperioides longipes (Chevreux, 1900) (cont.)

9   Pereiopod 1, with detail
10 Pereiopod 2
11 Pereiopod 4
12 Pereiopod 5, with detail

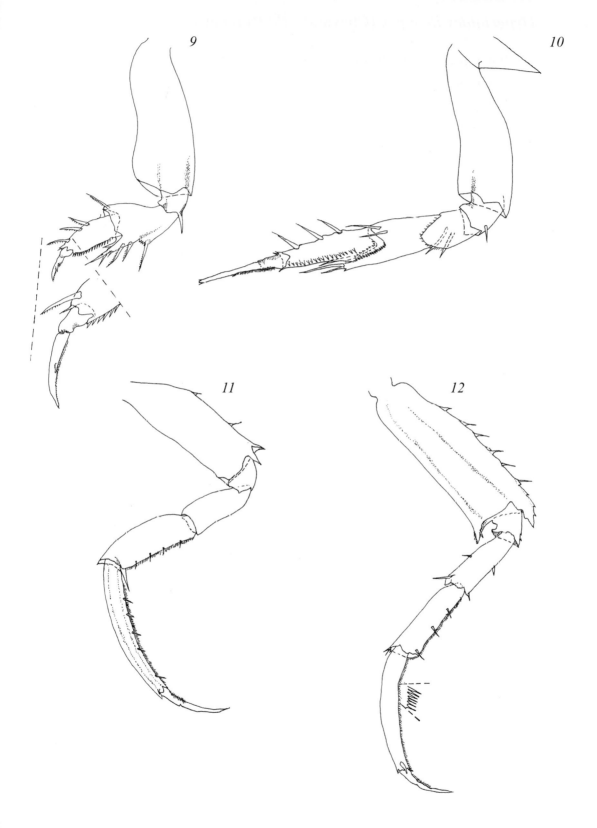

FIGURE 11c

Hyperioides longipes (Chevreux, 1900) (cont.)

13 Pereiopod 6, with details
14 Pereiopod 7
15 Pleonal epimeres
16 Uropods and telson

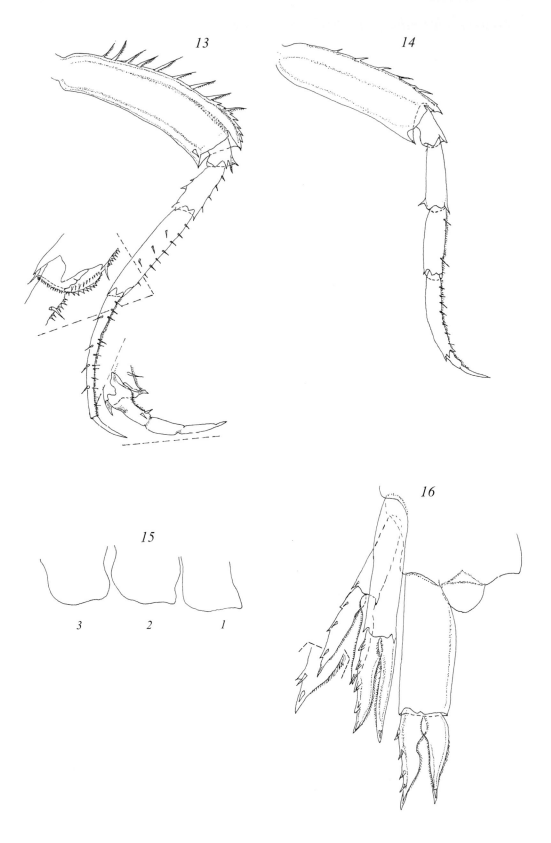

FIGURE 12a

Hyperioides sibaginis (Stebbing, 1888)

Female, 5 mm; male, 5 mm

1   Female, lateral view
2   Antennula, female
3   Antennula, male, lateral and ventral view
4   Antenna, female
5   Mandibula, female, molar plate
6   Mandibula, male, with palp

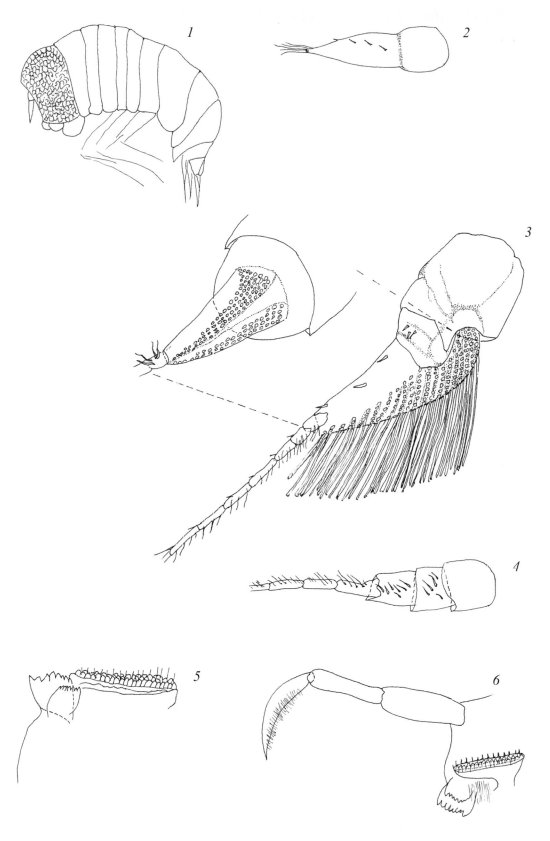

FIGURE 12b

Hyperioides sibaginis (Stebbing, 1888) (cont.)

7   Maxillula, female
8   Maxillula, male
9   Maxilla, female
10 Maxilla, male
11 Maxilliped, female
12 Pereiopod 1, female
13 Pereiopod 2, female
14 Pereiopod 3, female, with detail

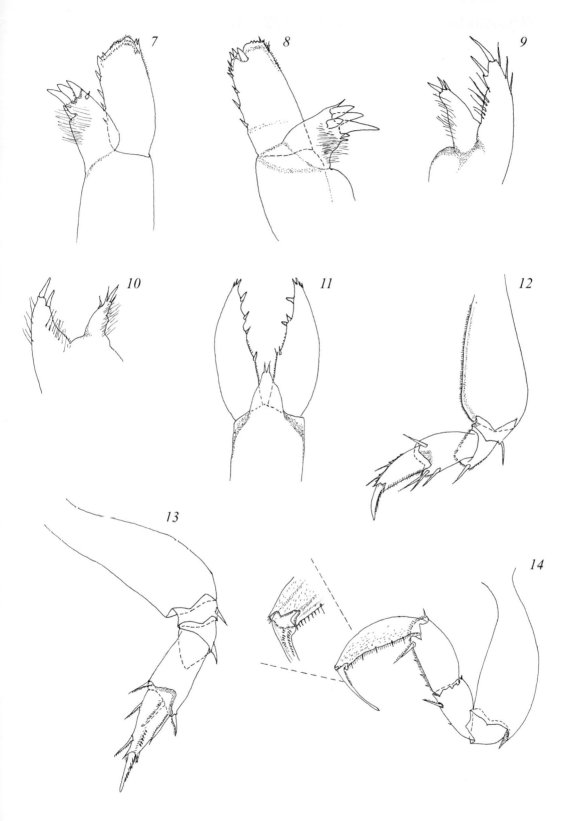

FIGURE 12c

Hyperioides sibaginis (Stebbing, 1888) (cont.)

15 Pereiopod 4, female
16 Pereiopod 5, female
17 Pereiopod 6, female
18 Pereiopod 6, male, with enlarged propus and dactylus
     and details
19 Pereiopod 7, female
20 Pleonal epimeres, female
21 Uropods and telson

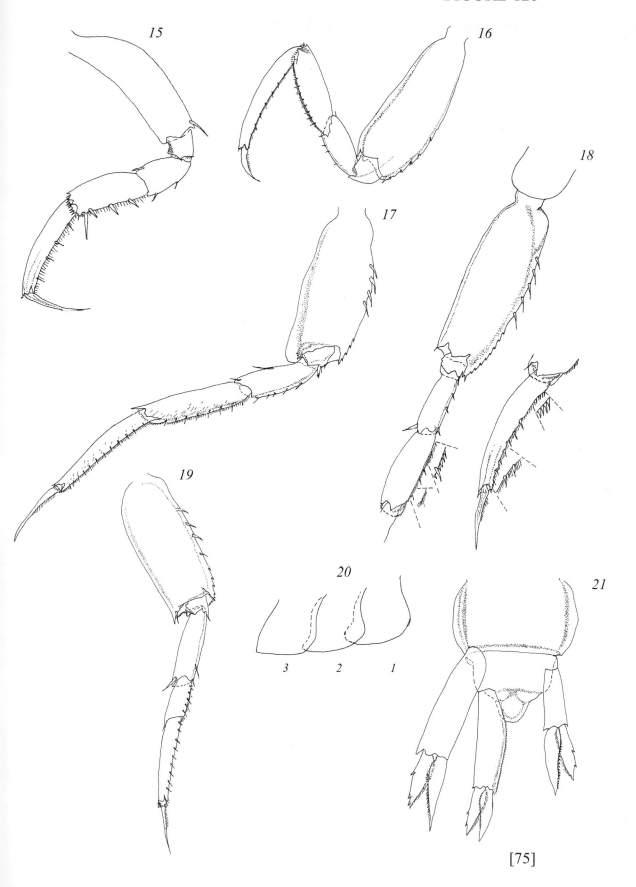

FIGURE 12c

15

16

17

18

19

20

3 2 1

21

[75]

FIGURE 13a

Lestrigonus schizogeneios (Stebbing, 1888)

Female, 2 mm; juvenile male, 2 mm

1   Male, lateral view
2   Female, lateral view, schematic
3   Antennula, male
4   Antenna, male
5   Mandibula, female
6   Mandibula, male

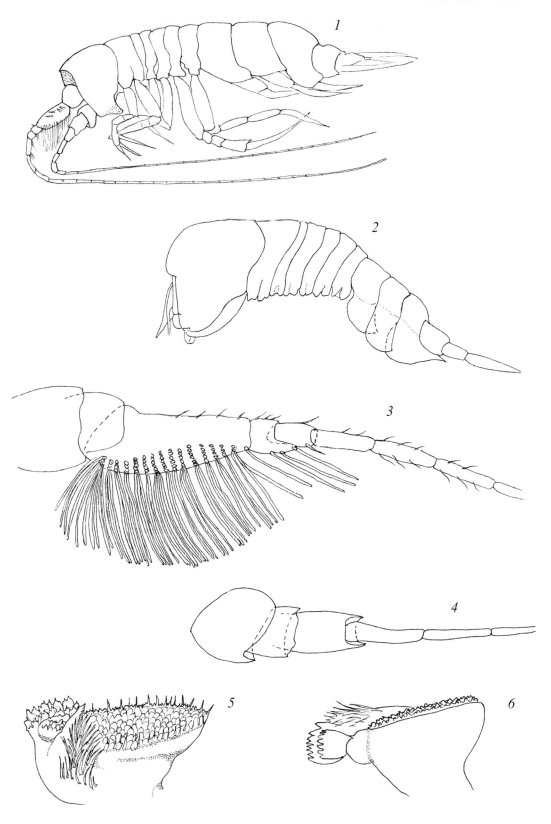

FIGURE 13b

Lestrigonus schizogeneios (Stebbing, 1888) (cont.)

7 Palp of mandibula, male
8 Maxillula, female
9 Maxillula, male
10 Maxilla, female
11 Maxilla, male
12 Maxilliped, male
13 Pereiopod 1, female
14 Pereiopod 1, male

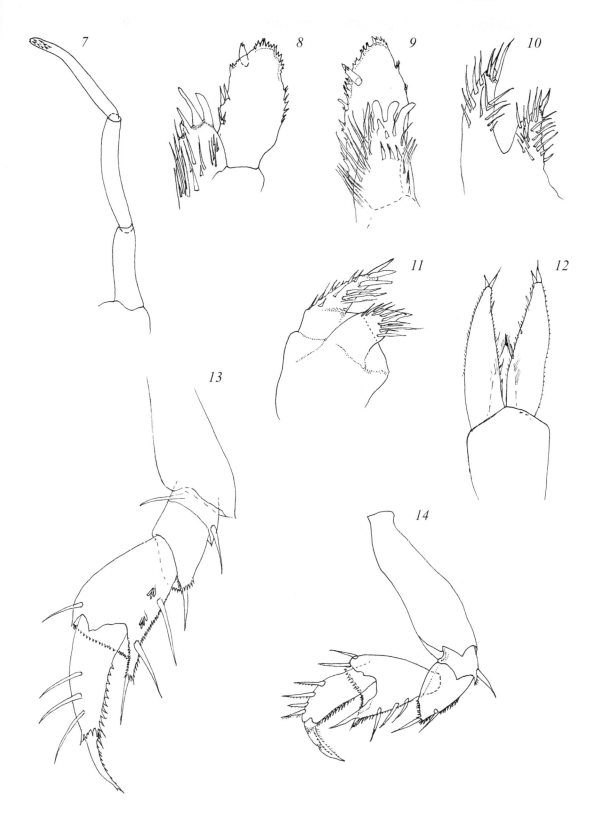

FIGURE 13c

Lestrigonus schizogeneios (Stebbing, 1888) (cont.)

15 Pereiopod 2, male
16 Pereiopod 3, male
17 Pereiopod 4, male
18 Pereiopod 5, male
19 Pereiopod 6, female
20 Pereiopod 6, male, poor armature type

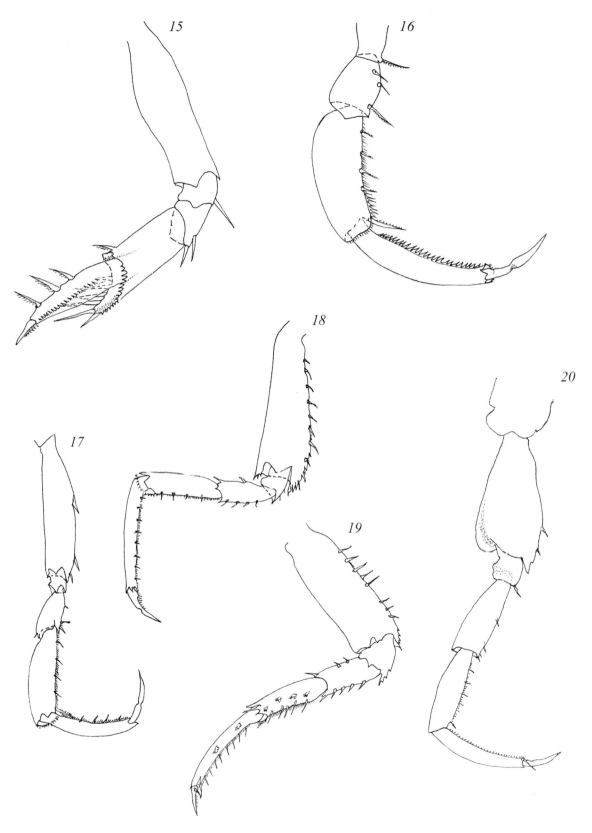

FIGURE 13d

Lestrigonus schizogeneios (Stebbing, 1888) (cont.)

21 Pereiopod 6, male, rich armature type
22 Pereiopod 7, male, poor armature type
23 Pereiopod 7, male, rich armature type
24 Last epimere, male
25 Uropods and telson, male

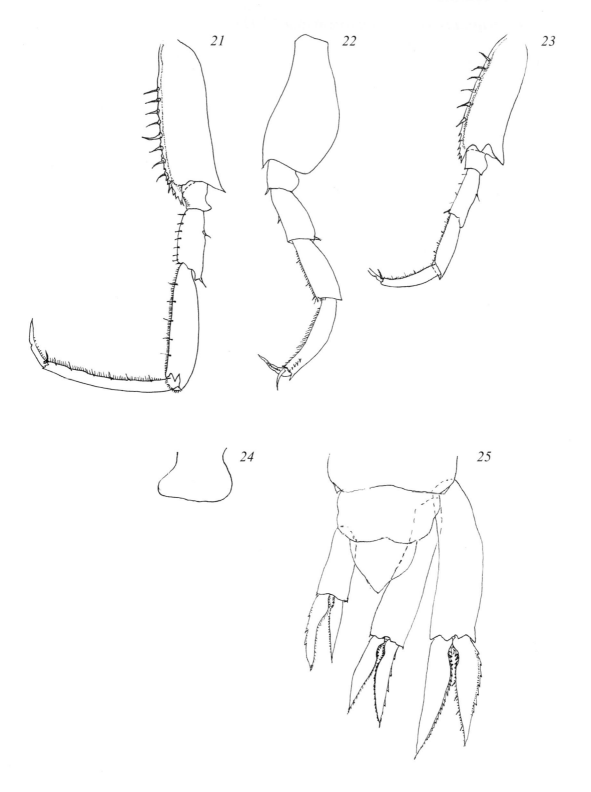

FIGURE 14a

Lestrigonus shoemakeri Bowman, 1973

Female, 4 mm; male, 3 mm

1   Female, lateral view, schematic
2   Male, lateral view, schematic
3   Female, dorsal view
4   Antennula, female
5   Mandibula, lateral and frontal views
6   Maxillula, female, with palp and outer lobe flattened
7   Maxilla
8   Maxilliped, female, lateral view

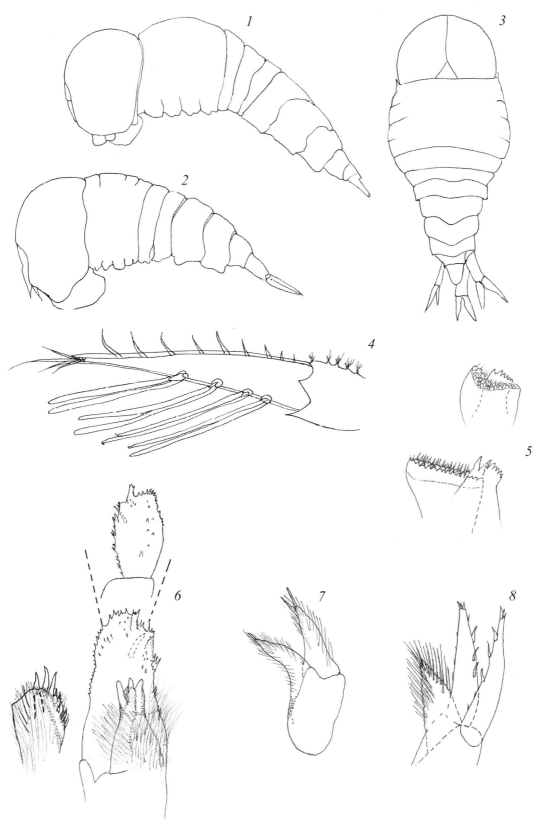

FIGURE 14b

Lestrigonus shoemakeri Bowman, 1973 (cont.)

9 Pereiopod 1, female
10 Pereiopod 1, male
11 Pereiopod 2, female
12 Pereiopod 2, male, with detail
13 Pereiopod 3, female
14 Pereiopod 3, male

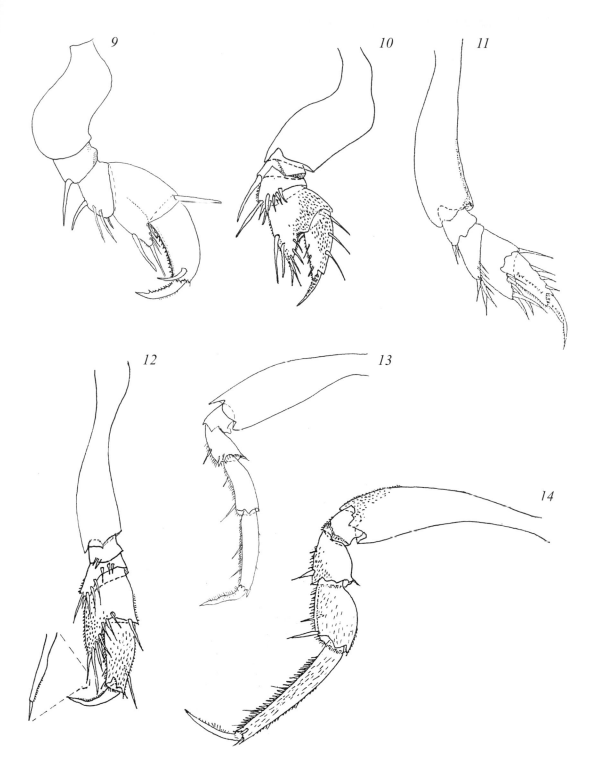

FIGURE 14c

Lestrigonus shoemakeri Bowman, 1973 (cont.)

15 Pereiopod 4, female
16 Pereiopod 5, female, with detail
17 Pereiopod 6, female, with detail
18 Pereiopod 7, female, with detail
19 Uropods and telson

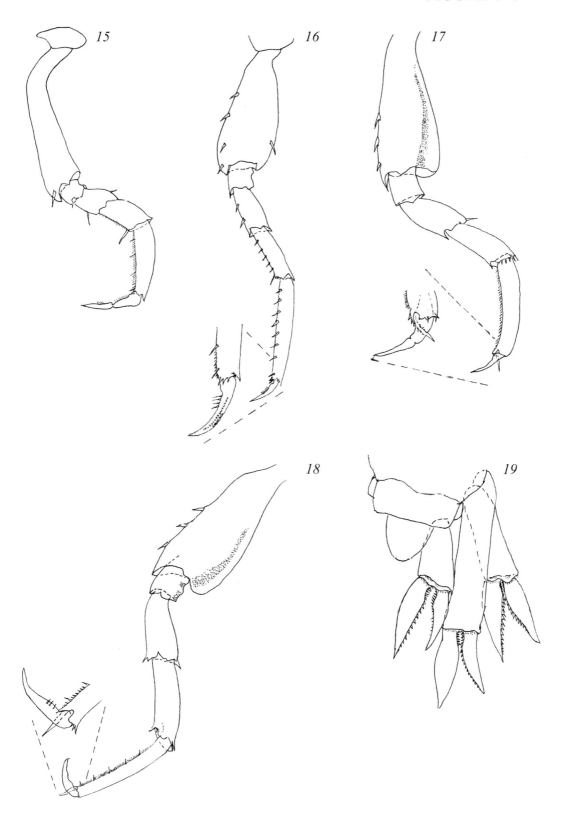

FIGURE 15a

Lestrigonus macrophthalmus (Vosseler, 1901)

Female, 3 mm

1   Mandibula
2   Maxillula
3   Maxilla
4   Maxilliped
5   Pereiopod 1
6   Pereiopod 2, with detail
7   Pereiopod 3

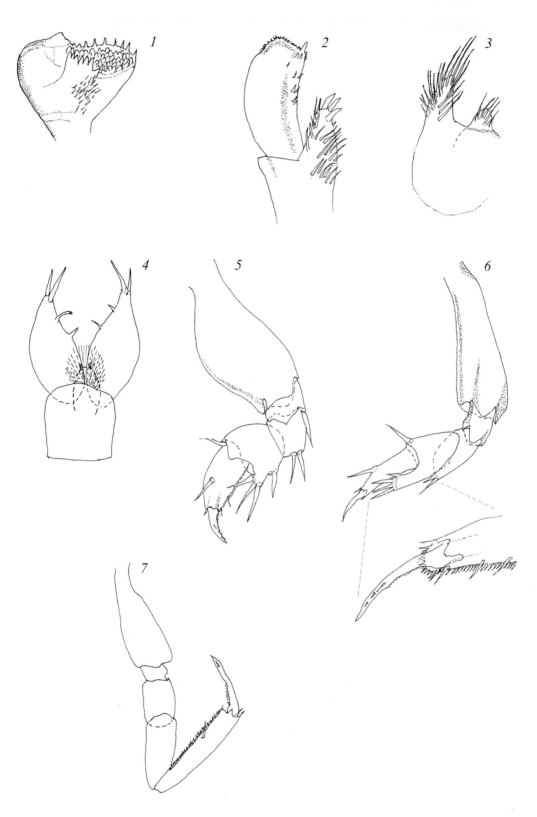

FIGURE 15b

Lestrigonus macrophthalmus (Vosseler, 1901) (cont.)

8 Pereiopod 5, with detail
9 Pereiopod 6, part of propus and dactylus
10 Pereiopod 7
11 Uropods and telson

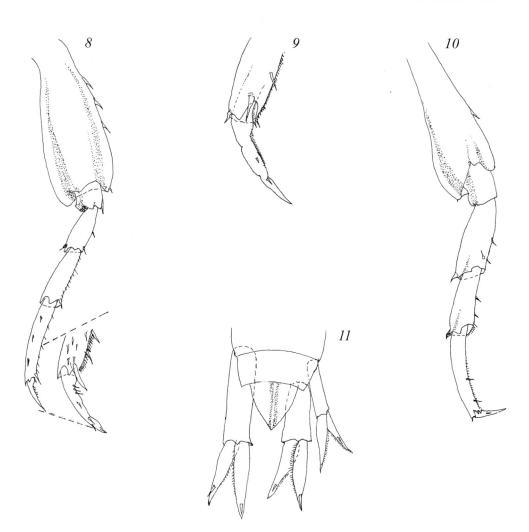

FIGURE 16a

Lestrigonus bengalensis Giles, 1887

Females and males, 3–4 mm

1   Lateral view, male
2   Antennula, female
3   Antennula, male
4   Mandibula, female
5   Mandibula, male
6   Maxillula, female
7   Maxillula, another view
8   Maxilla, female
9   Maxilla, another view

FIGURE 16a

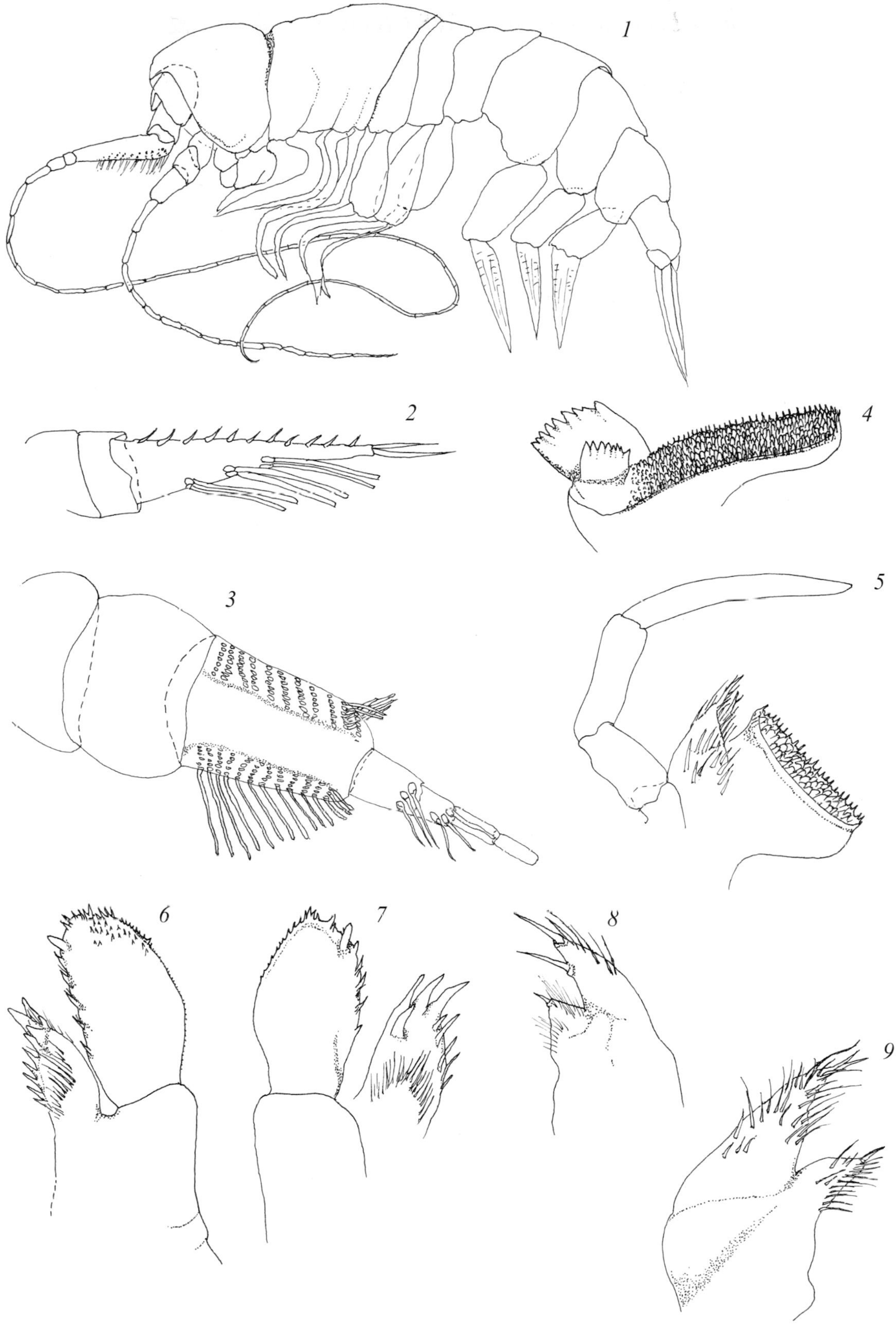

FIGURE 16b

Lestrigonus bengalensis Giles, 1887 (cont.)

10 Maxilliped, male
11 Pereiopod 1, female
12 Pereiopod 1, male
13 Pereiopod 2, female
14 Pereiopod 2, male
15 Pereiopod 3, female
16 Pereiopod 3, male, with epipodite

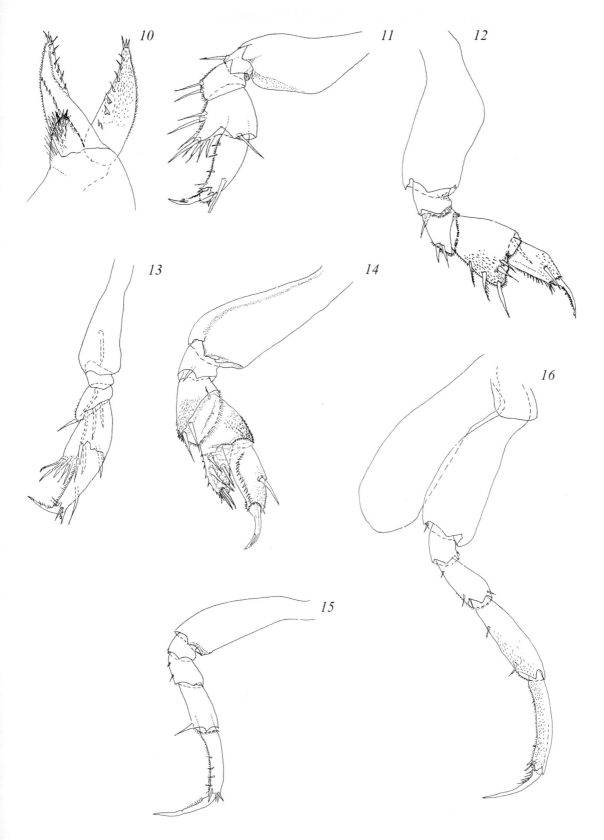

FIGURE 16b

10

11

12

13

14

16

15

[97]

FIGURE 16c

Lestrigonus bengalensis Giles, 1887 (cont.)

17 Pereiopod 4, female, with detail
18 Pereiopod 4, male
19 Pereiopod 5, female, with detail
20 Pereiopod 5, male, with detail
21 Pereiopod 6, female

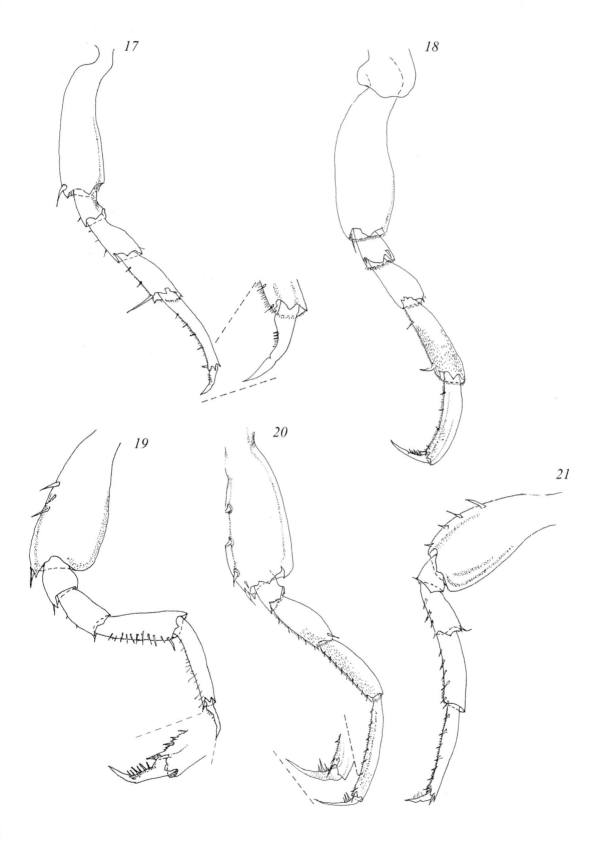

FIGURE 16d

Lestrigonus bengalensis Giles, 1887 (cont.)

22 Pereiopod 6, male
23 Pereiopod 7, male
24 Uropods and telson, male
25 Armature of uropods 1, 2 and 3, female

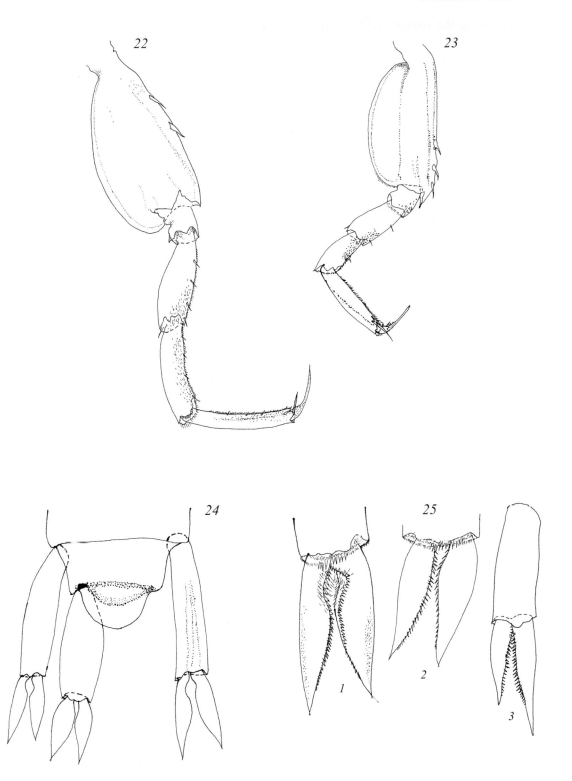

FIGURE 17a

Hyperietta luzoni (Stebbing, 1888)

Female, 2 mm; male, 4 mm

1   Female lateral view, schematic
2   Mandibula, female
3   Maxillula, female
4   Maxilla, female
5   Maxilliped, female
6   Maxilliped, male
7   Pereiopod 1
8   Pereiopod 2
9   Pereiopod 4

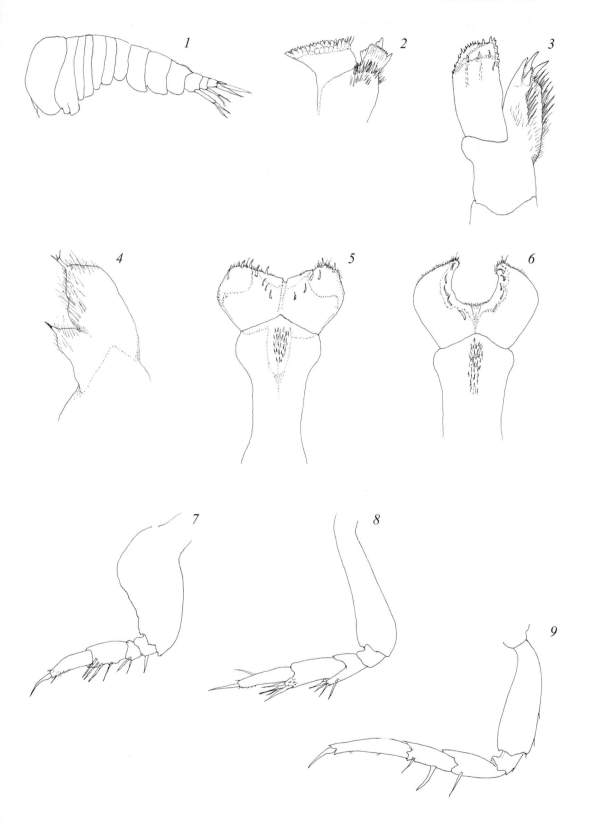

FIGURE 17b

Hyperietta luzoni (Stebbing, 1888) (cont.)

10 Pereiopod 5, with detail
11 Pereiopod 6, with detail
12 Pereiopod 7, with detail
13 Bifid spines and coupling hooks of pleopods
14 Uropods 1, 2 and 3

FIGURE 17b

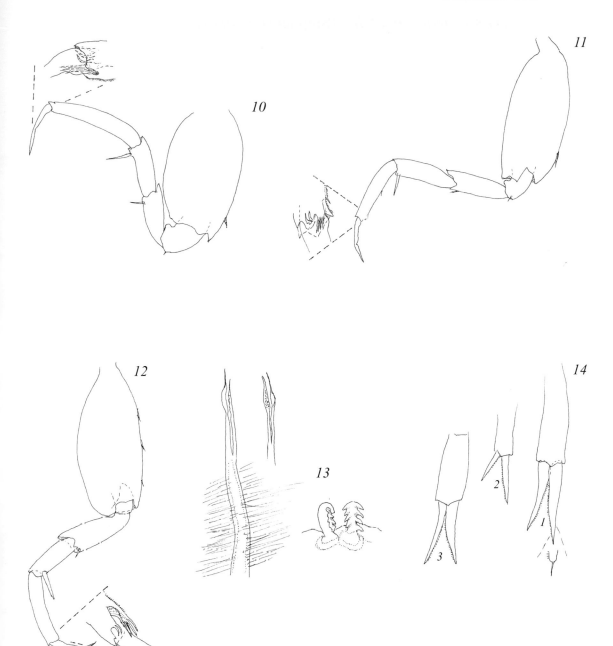

FIGURE 18

Hyperionyx macrodactylus (Stephensen, 1924)

Female, around 3 mm

1   Lateral view, schematic
2   Maxillula
3   Maxilla
4   Maxilliped, posterior view
5   Pereiopod 1 with typically bifurcate spine
6   Pereiopod 2
7   Pereiopod 3
8   Pereiopod 4
9   Pereiopod 5
10 Pereiopod 6
11 Pereiopod 7
12 Bifid spine and coupling hooks of pleopods
13 Uropods and telson

FIGURE 18

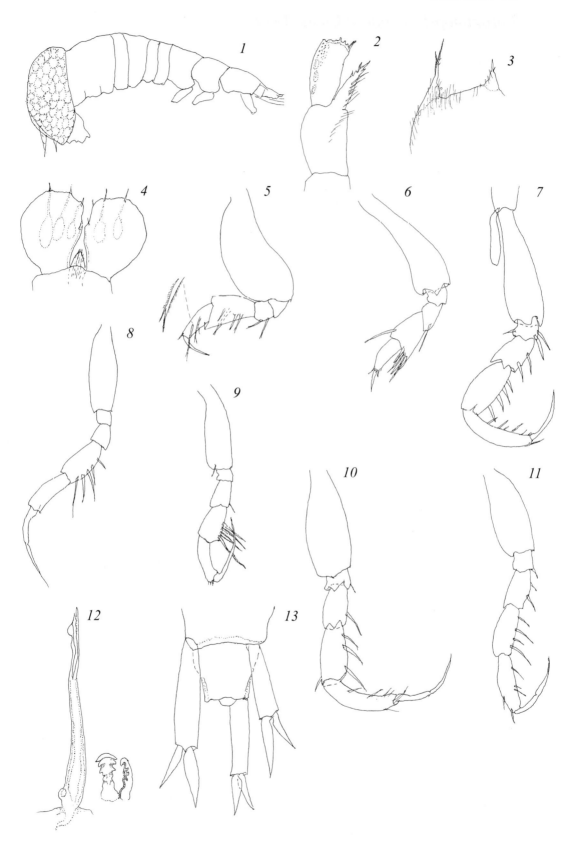

FIGURE 19a

Phronimopsis spinifera Claus, 1879

Females, 6 and 7 mm

1 Lateral view
2 Antennula
3 Antenna
4 Mandibula
5 Mandibula, enlarged
6 Maxillula
7 Maxilla

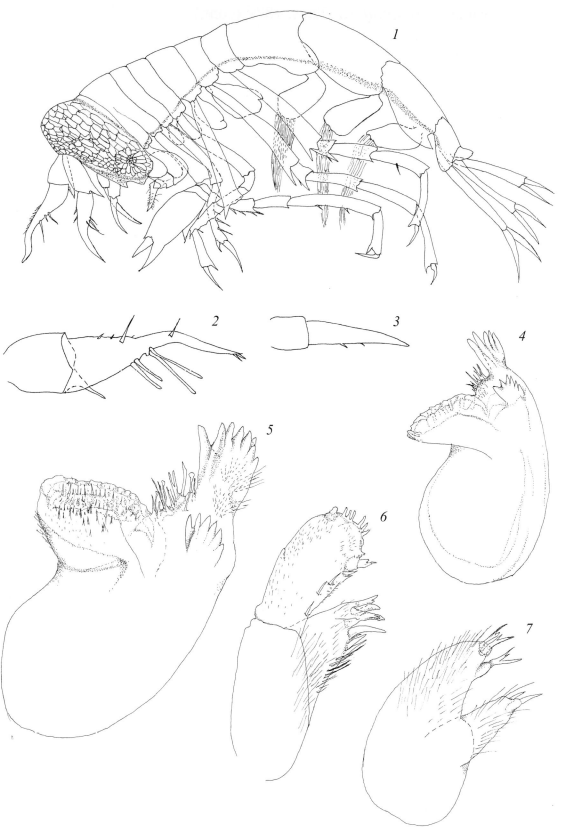

FIGURE 19b

Phronimopsis spinifera Claus, 1879 (cont.)

8   Maxillula, palp
9   Maxillula, outer lobe
10 Maxilliped, lateral view
11 Maxilliped, antero-lateral view
12 Maxilliped, inner lobe, flattened
13 Maxilliped, apex of outer lobe
14 Pereiopod 1
15 Pereiopod 2
16 Pereiopod 3

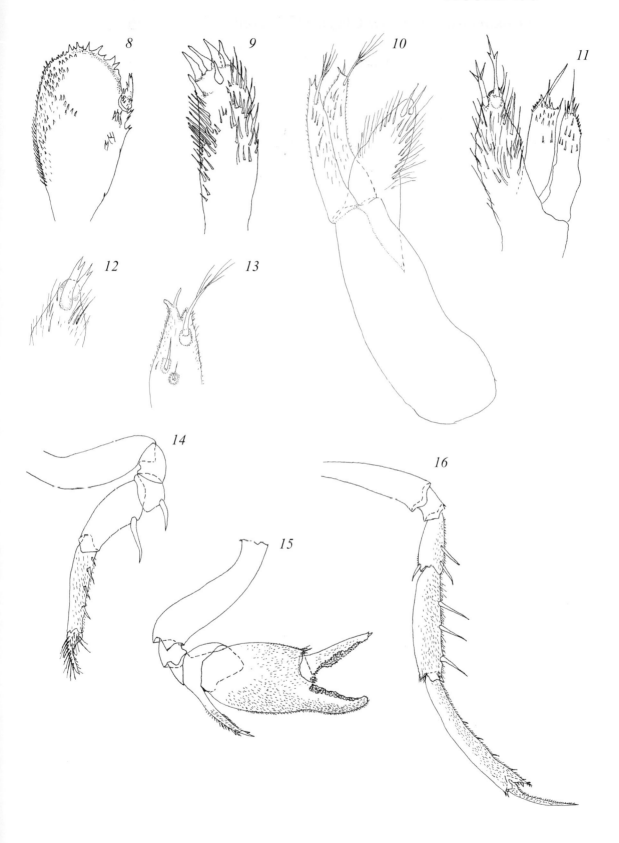

FIGURE 19c

Phronimopsis spinifera Claus, 1879 (cont.)

17 Pereiopod 4, with detail
18 Pereiopod 5
19 Pereiopod 7
20 Bifid spine and coupling hooks of pleopod 1
21 Bifid spine and coupling hooks of pleopod 2
22 Uropods and telson

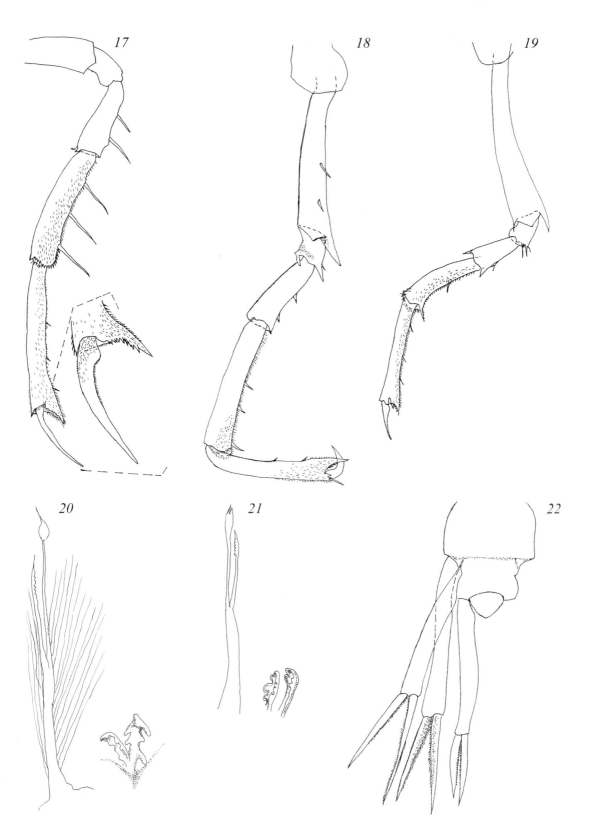

FIGURE 20a

PHRONIMIDAE Dana

Phronima sedentaria Forskål, 1775

Size 1: 0.5 mm, sex undefined; Size 2: 5 mm, female; Size 3: 12 mm, female and male; Size 4: 28 mm, female; Size 5: 40 mm, female

1   Lateral view, female size 3
2   Lateral view, male size 3
3   Antennula with aesthetascs, female size 2

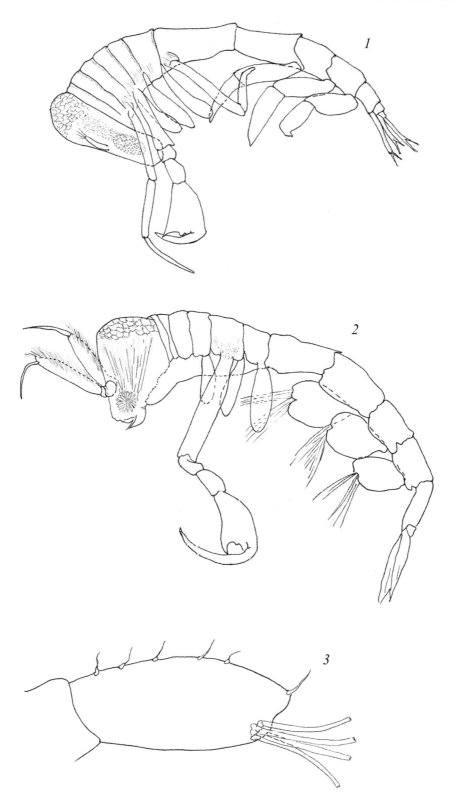

FIGURE 20b

Phronima sedentaria Forskål, 1775 (cont.)

4   Antennula, male size 4, with details
5   Antenna, female size 5
6   Mandibula, specimen size 1
7   Mandibula, female size 2
8   Mandibula, female size 4
9   Maxillula, female size 2
10 Maxillula, female size 4

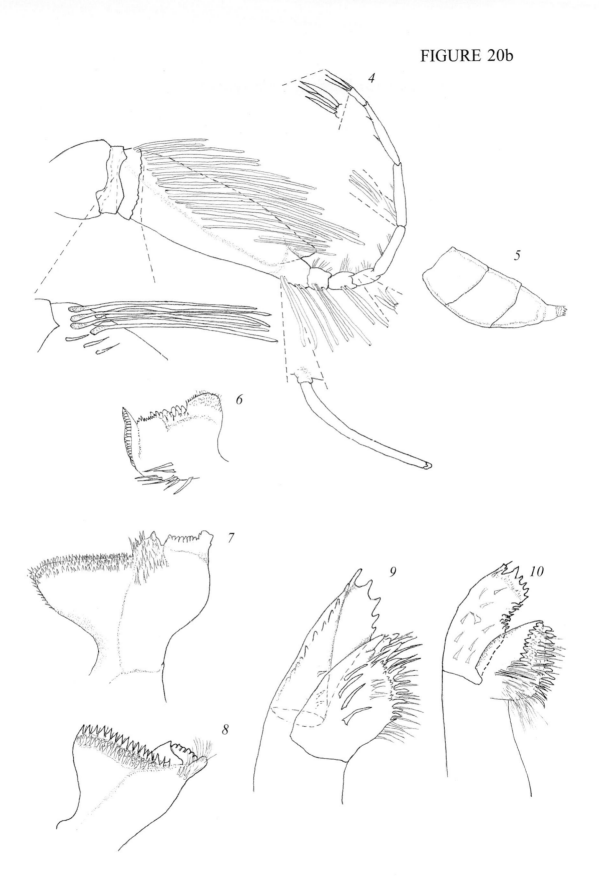

FIGURE 20c

Phronima sedentaria Forskål, 1775 (cont.)

11 Maxilla, female size 2
12 Maxilla, female size 4
13 Maxilliped, female size 4
14 Pereiopod 1, specimen size 1, with detail of dactylus
15 Pereiopod 1, female size 2, with detail of dactylus

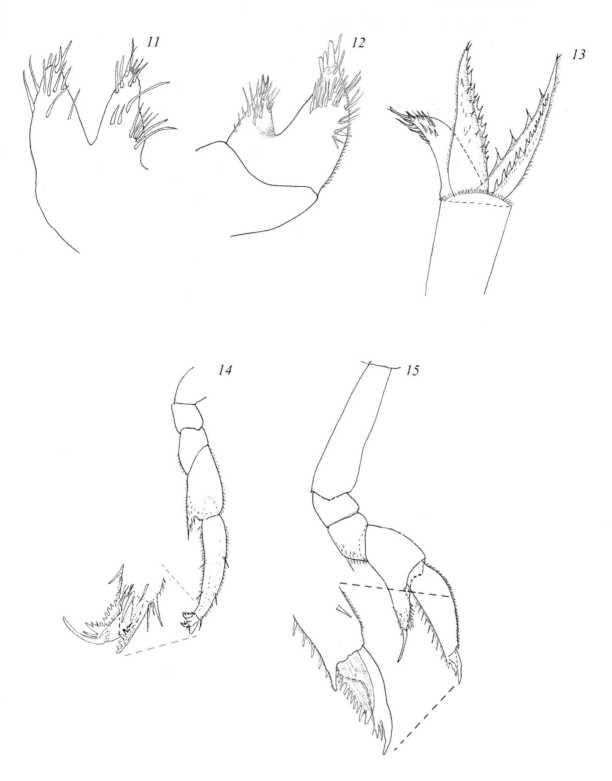

FIGURE 20d

Phronima sedentaria Forskål, 1775 (cont.)

16 Pereiopod 1, specimen size 3, with details of dactylus
17 Pereiopod 2, fragment, specimen size 1, with detail of dactylus
18 Pereiopod 2, female size 2, with detail of dactylus
19 Pereiopod 2, male size 3, with detail of dactylus

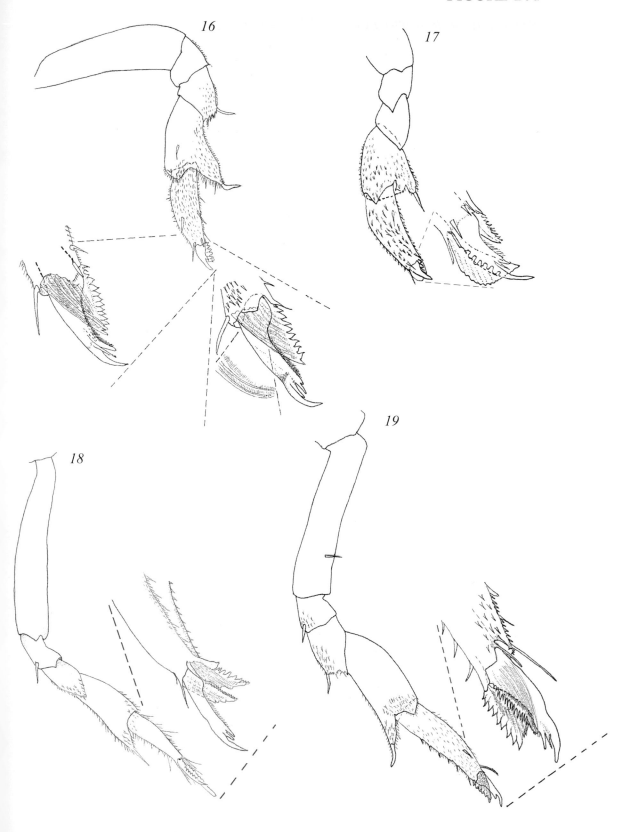

FIGURE 20e

Phronima sedentaria Forskål, 1775 (cont.)

20 Pereiopod 2, female size 4
21 Pereiopod 2, female size 4, dactylus
22 Pereiopod 2, female size 5
23 Pereiopod 3, specimen size 1, distal part of propus with dactylus
24 Pereiopod 3, specimen size 3, with detail

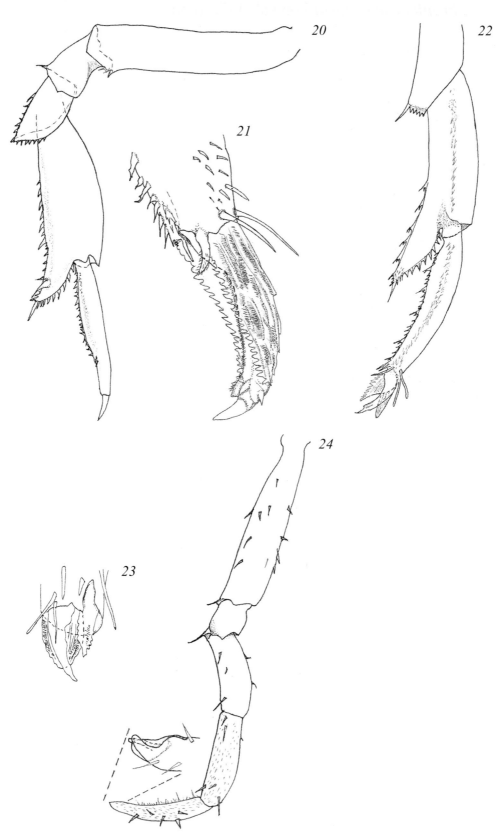

FIGURE 20f

Phronima sedentaria Forskål, 1775 (cont.)

25 Pereiopod 4, specimen size 3, with detail
26 Pereiopods 5, left and right, specimen size 1
    (note left–right asymmetry)
27 Pereiopod 5, female size 2, with detail
28 Pereiopod 5, female size 3

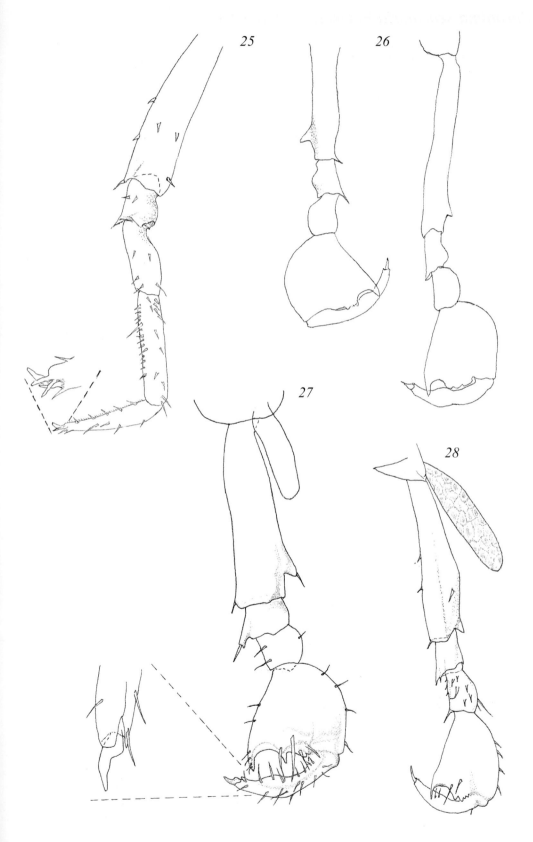

FIGURE 20g

Phronima sedentaria Forskål, 1775 (cont.)

29 Pereiopod 5, male size 3, with detail
30 Pereiopod 5, male size 3, subchela with detail
31 Pereiopod 5, female size 4
32 Pereiopod 5, female size 5, tip of subchela

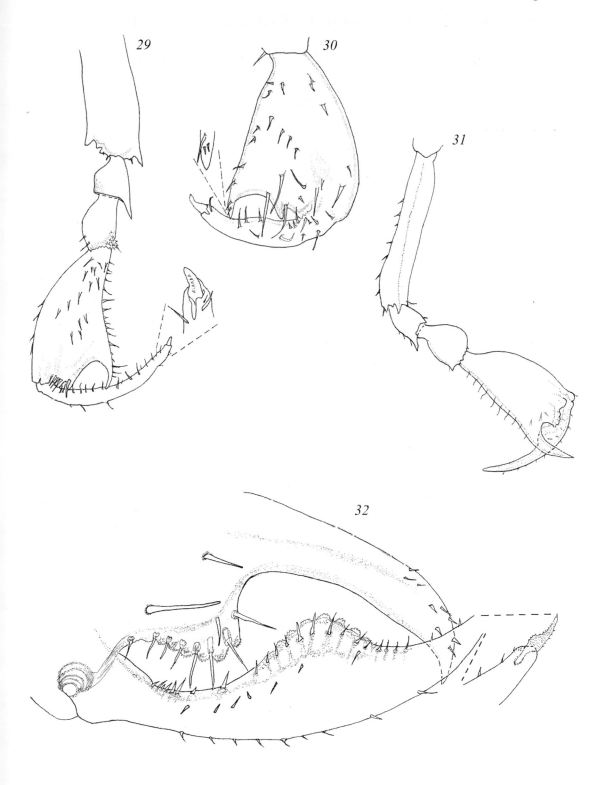

FIGURE 20h

Phronima sedentaria Forskål, 1775 (cont.)

33 Pereiopod 6, specimen size 3, with detail
34 Pereiopod 7, specimen size 3, with detaill
35 Uropods and telson, female size 2
36 Uropods and telson, female size 3

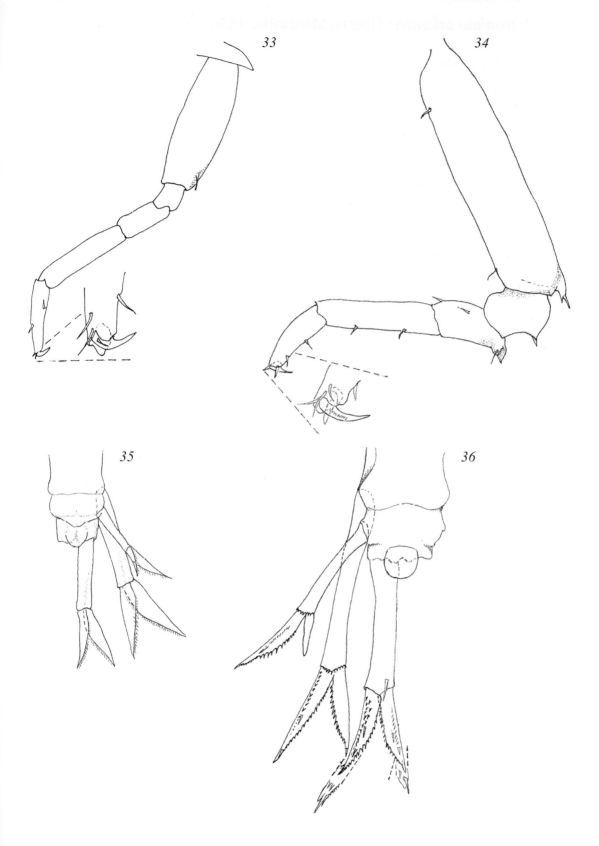

FIGURE 21a

Phronima atlantica Guérin-Ménéville, 1836

Female, 9 mm

1   Antennula
2   Mandibula, molar process flattened
3   Maxillula, outer lobe
4   Maxillula, inner lobe flattened
5   Maxilla
6   Maxilliped, ventrolateral view

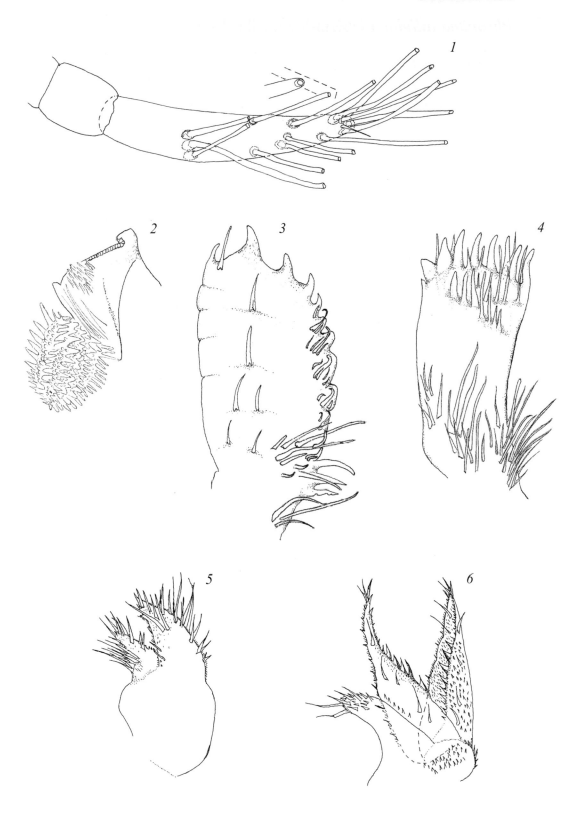

FIGURE 21b

Phronima atlantica Guérin-Ménéville, 1836 (cont.)

7   Pereiopod 1
8   Pereiopod 1, dactylus
9   Pereiopod 3, with detail
10 Pereiopod 4, with detail

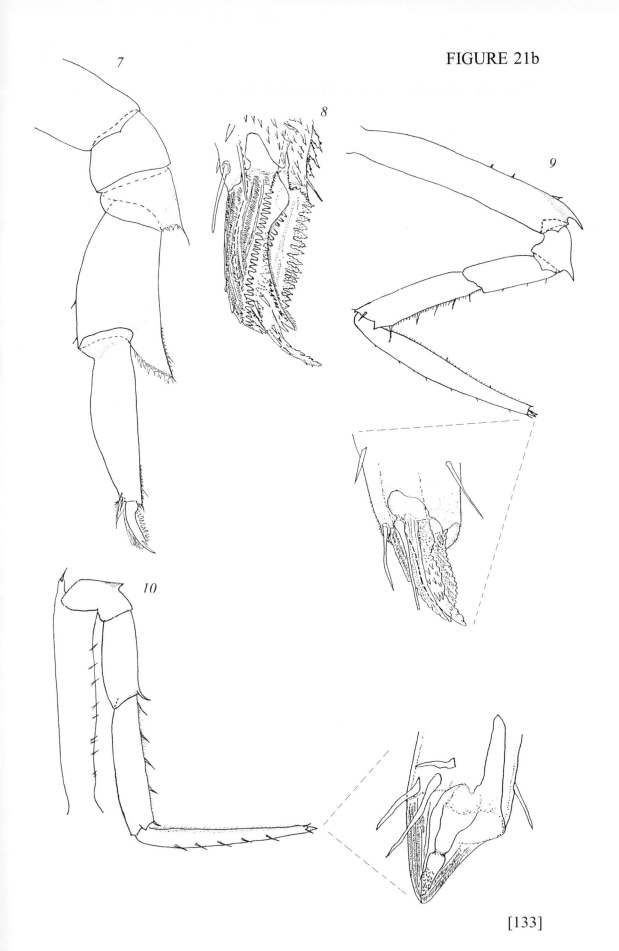

FIGURE 21c

Phronima atlantica Guérin-Ménéville, 1836 (cont.)

11 Pereiopod 5
12 Pereiopod 5, tip of subchela
13 Pereiopod 6, dactylus
14 Pereiopod 7, dactylus
15 Bifid spine on pleopod 1
16 Bifid spines on pleopod 2
17 Coupling hooks of pleopods
18 Uropods, with detail and telson

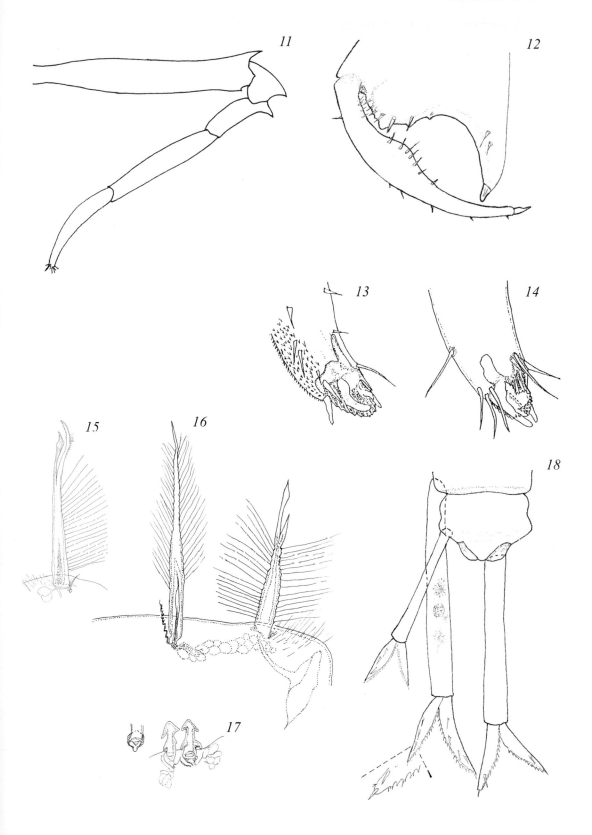

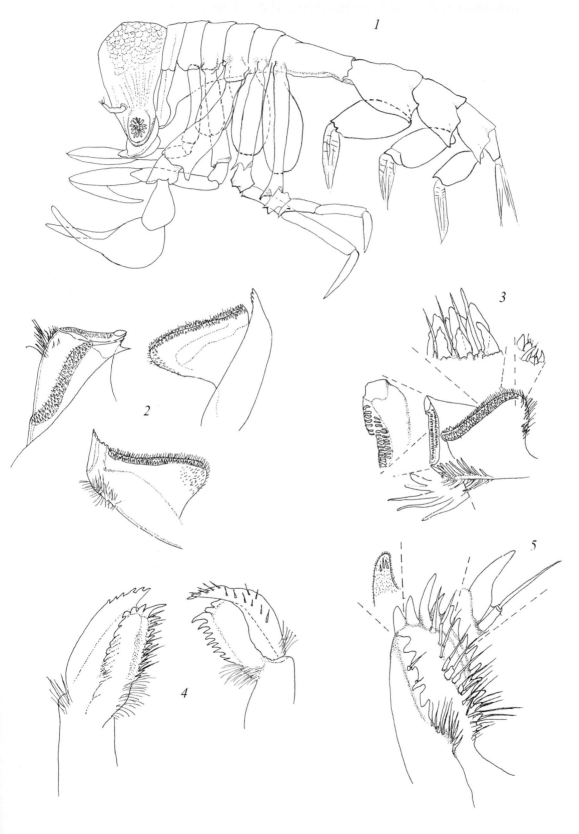

FIGURE 22b

Phronima solitaria Guérin-Ménéville, 1836 (cont.)

6 Maxillula, outer lobe, flattened and relatively enlarged
7 Maxilla
8 Maxilla, tips of inner and outer lobes
9 Maxilliped, anterior view
10 Maxilliped, tip of outer lobe
11 Maxilliped, lateral view
12 Maxilliped, inner lobe

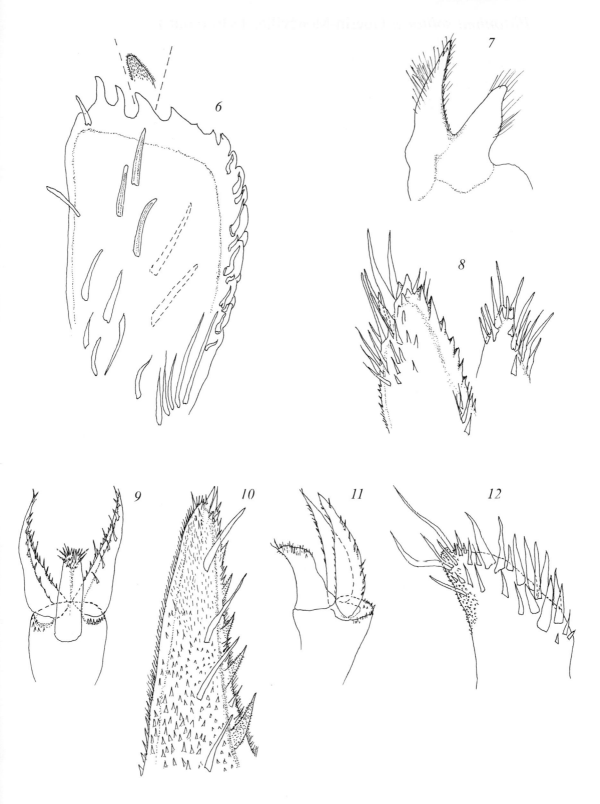

FIGURE 22c

Phronima solitaria Guérin-Ménéville, 1836 (cont.)

13 Pereiopod 1, with details
14 Pereiopod 1, propus and dactylus,
     with details of costate lamellae
15 Pereiopod 1, 26 mm specimen, propus with
     costate lamellae and dactylus

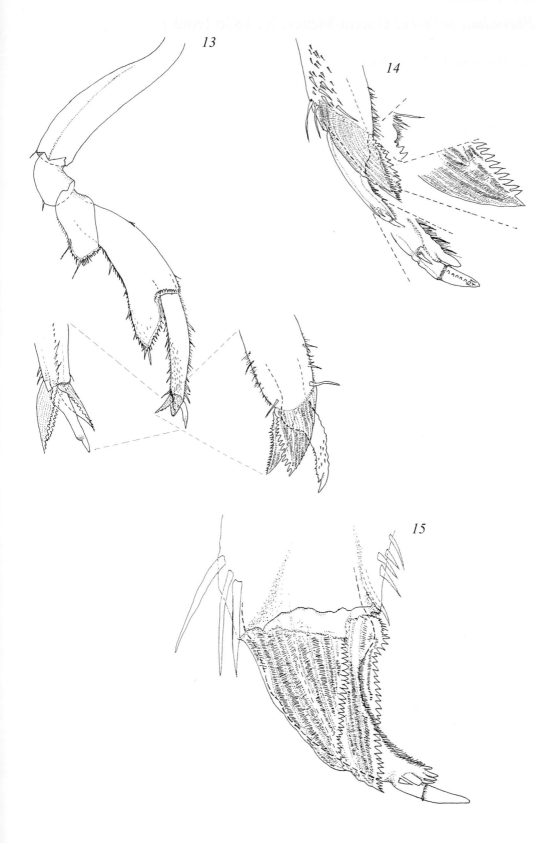

FIGURE 22d

Phronima solitaria Guérin-Ménéville, 1836 (cont.)

16 Pereiopod 2, with details
17 Pereiopod 2, distal part of propus with costate lamellae
    and dactylus
18 Pereiopod 4
19 Pereiopod 5, with subchela magnified

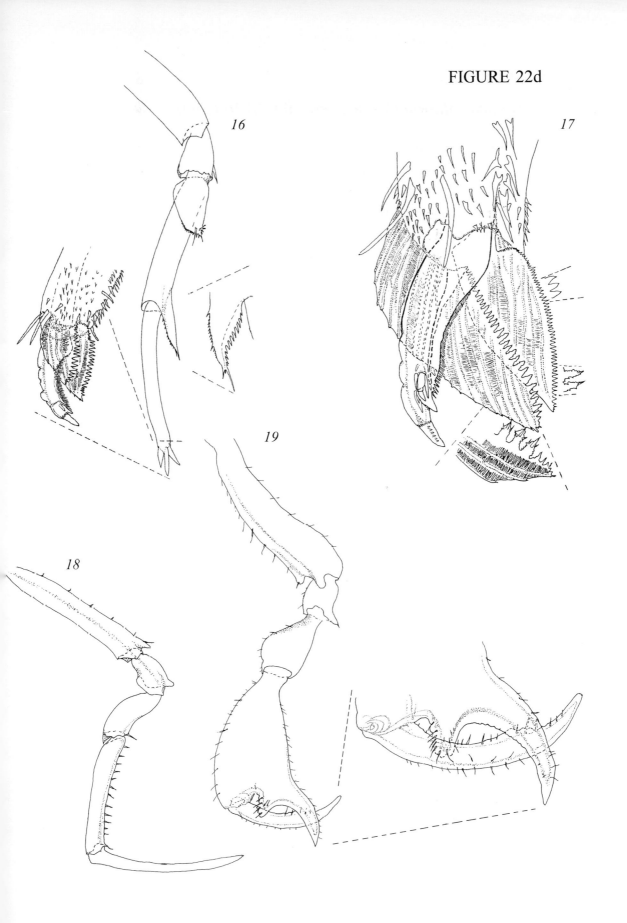

FIGURE 22e

Phronima solitaria Guérin-Ménéville, 1836 (cont.)

20 Pereiopod 6, dactylus
21 Pereiopod 7, dactylus
22 Ventral edge of last epimere
23 First bifid spine on pleopod 1
24 Second bifid spine on pleopod 1
25 Bases of first and second bifid spines on pleopod 1
26 First pair of coupling hooks on pleopod 1
27 Second pair of coupling hooks on pleopod 1
28 Uropods and telson

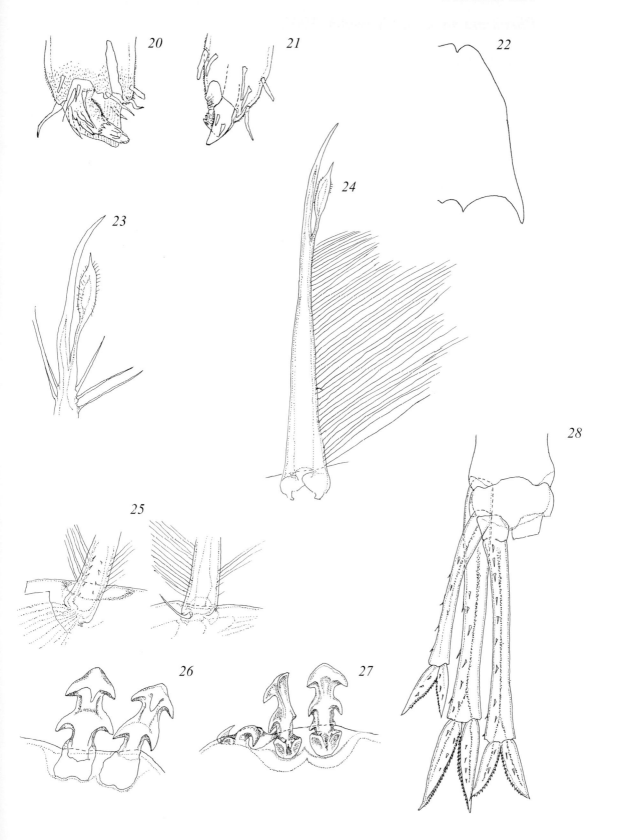

FIGURE 22e

FIGURE 23a

Phronima stebbingi Vosseler, 1901

Females and males, 6 mm

1   Lateral view, female
2   Lateral view, male
3   Antennula, male
4   Antennula, female
5   Antenna, male
6   Mandibula, female
7   Maxillula, female, inner and outer lobes separate

FIGURE 23a

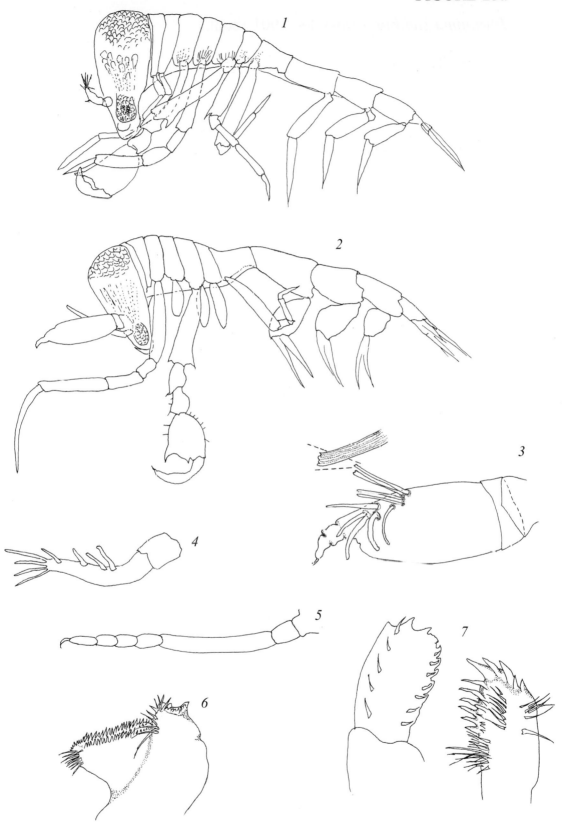

[147]

FIGURE 23b

Phronima stebbingi Vosseler, 1901 (cont.)

8   Maxilla
9   Maxilliped, female
10 Pereiopod 1, female, with details of dactylus and propus
11 Pereiopod 1, male, with detail
12 Pereiopod 2, female, with detail

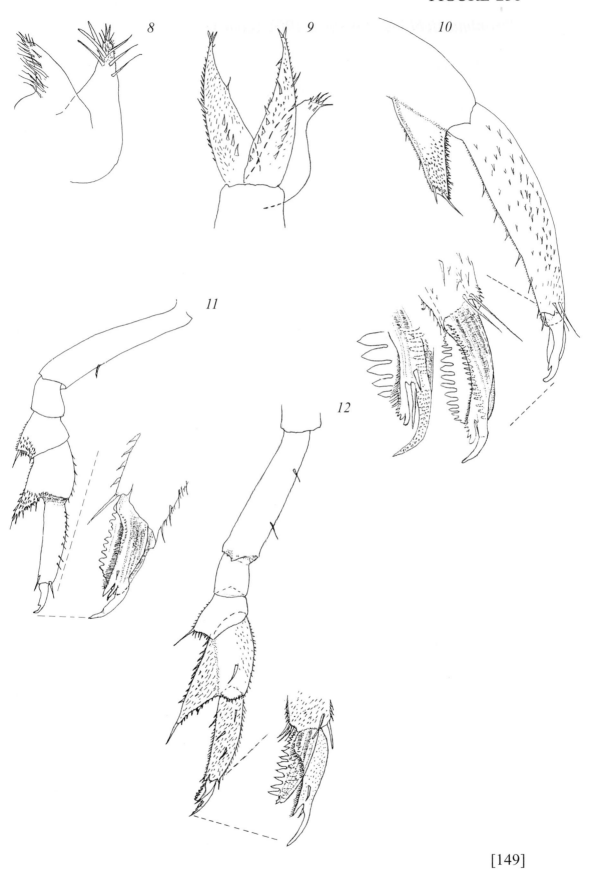

FIGURE 23c

Phronima stebbingi Vosseler, 1901 (cont.)

13 Pereiopod 2, male, with details
14 Pereiopod 3, female, with detail
15 Pereiopod 4, female
16 Pereiopod 5, female, with details

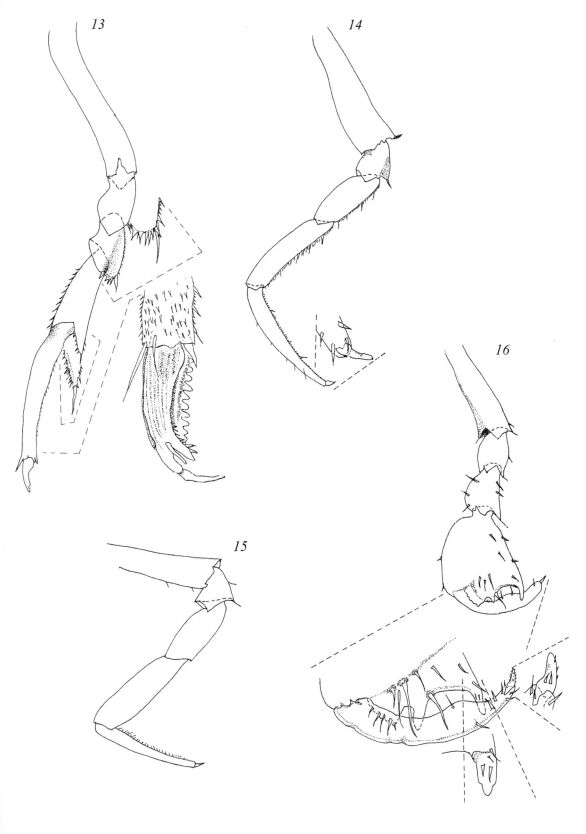

FIGURE 23d

Phronima stebbingi Vosseler, 1901 (cont.)

17 Pereiopod 5, male, with detail
18 Pereiopod 6, female, with detail
19 Pereiopod 7, female, with detail
20 Uropods and telson, female, with detail
21 Uropods and telson, male

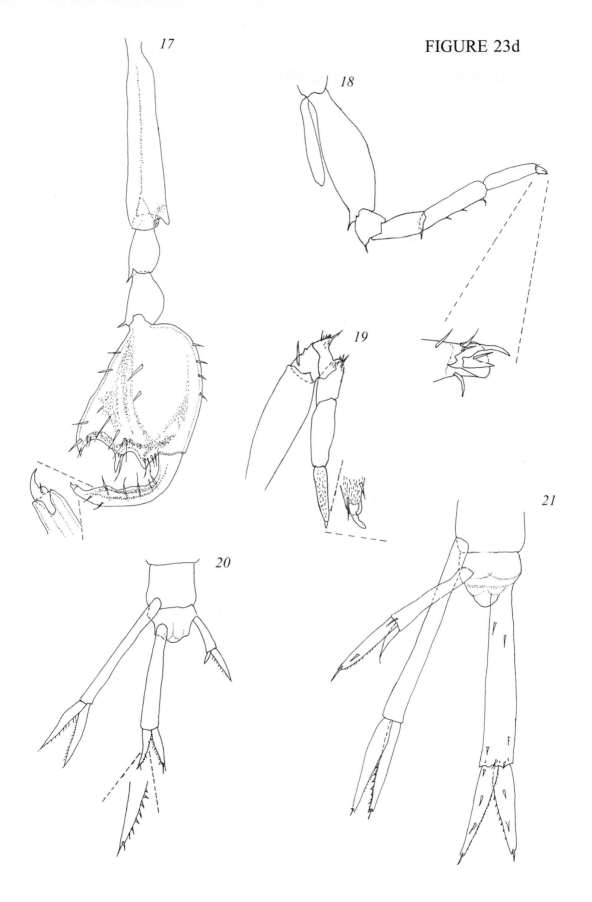

FIGURE 23d

FIGURE 24a

Phronima curvipes Vosseler, 1901

Female, 8 mm; male, 10 mm

1   Lateral view (pereiopod 3 ommitted)
2   Antennula, male
3   Maxilliped, female
4   Pereiopod 1, female
5   Pereiopod 2, female, with detail

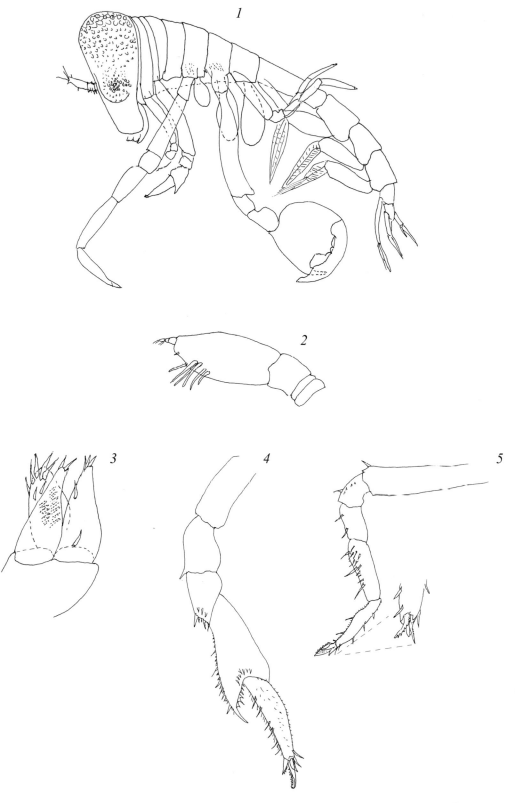

FIGURE 24b

Phronima curvipes Vosseler, 1901 (cont.)

6   Pereiopod 3, female
7   Pereiopod 5, female
8   Pereiopod 5, male
9   Pereiopod 7, female
10 Uropods and telson, female

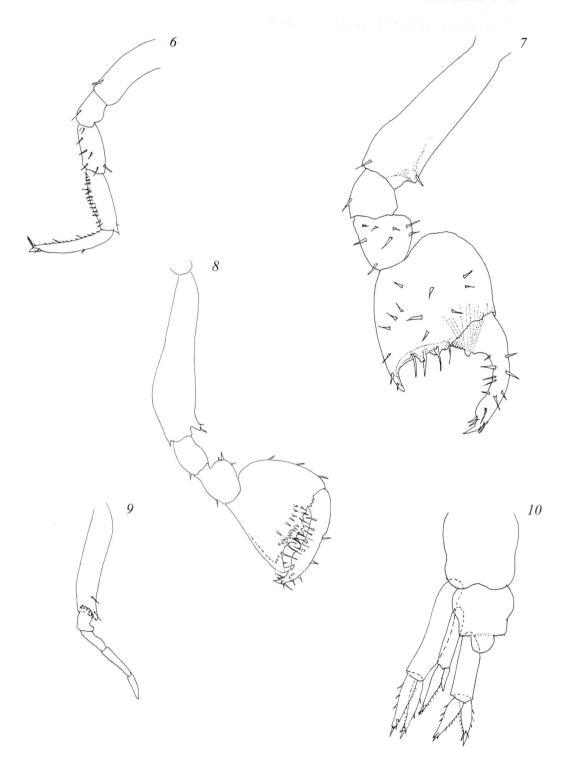

FIGURE 25a

Phronima colletti Bovallius, 1887

Females, 8 and 9 mm

1 Lateral view
2 Antennula
3 Mandibula
4 Maxillula
5 Maxilla
6 Maxilliped

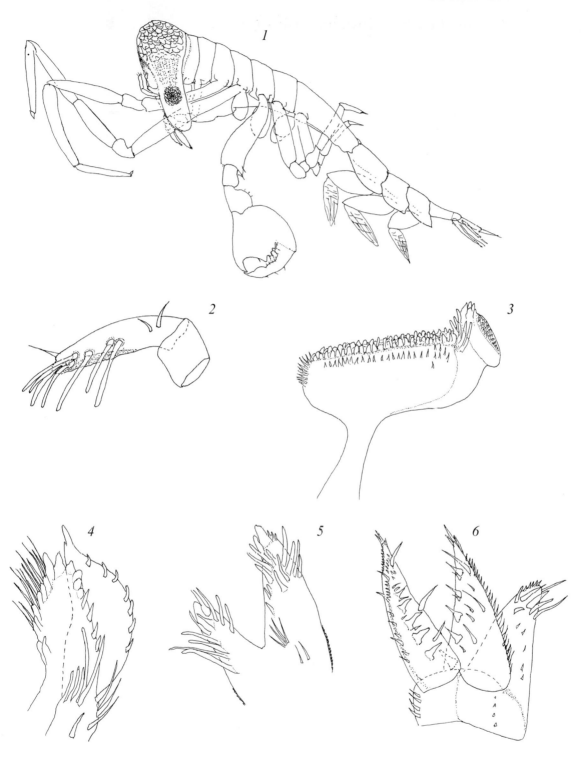

FIGURE 25b

Phronima colletti Bovallius, 1887 (cont.)

7   Pereiopod 1, with detail
8   Pereiopod 2, with detail
9   Pereiopod 3
10 Pereiopod 4

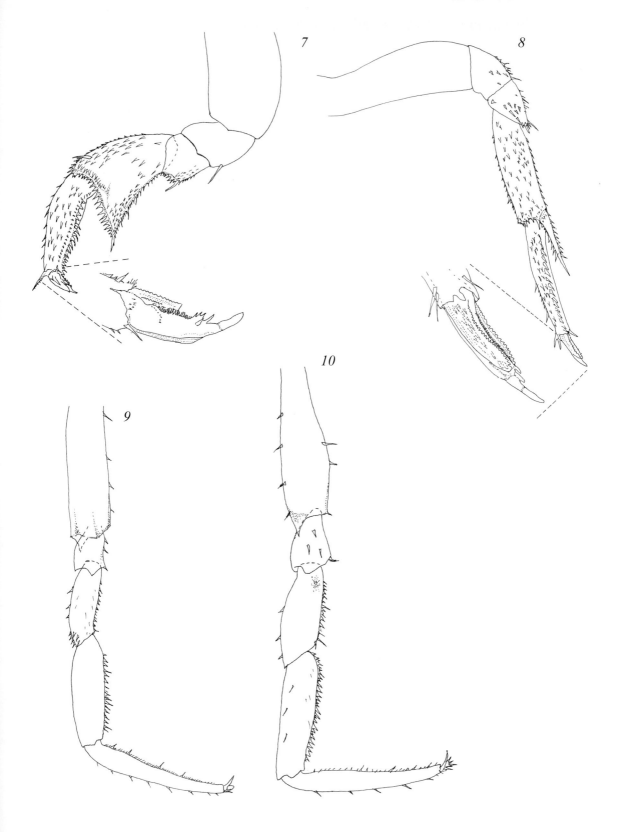

FIGURE 25c

Phronima colletti Bovallius, 1887 (cont.)

11 Pereiopod 5
12 Pereiopod 5, detail of subchela
13 Pereiopod 6, with detail
14 Pereiopod 7, with detail
15 Uropods and telson

FIGURE 25c

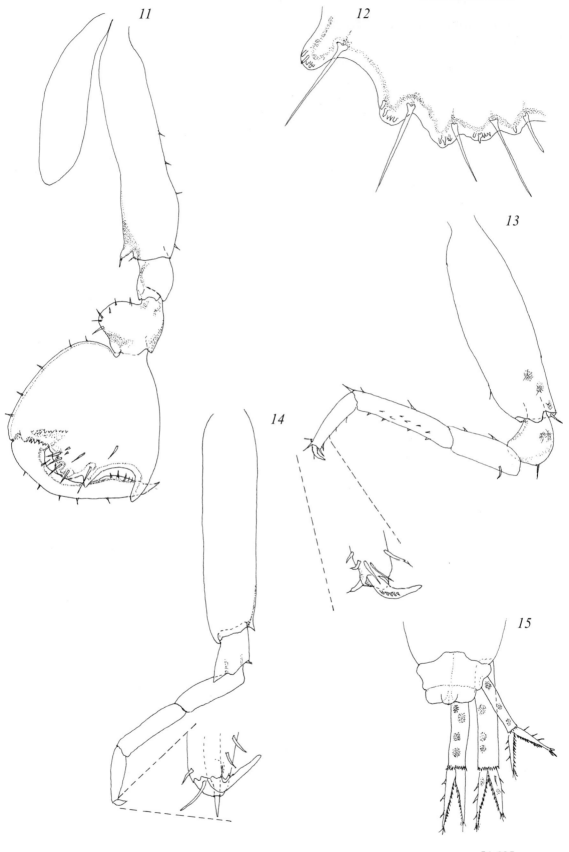

FIGURE 26a

Phronima pacifica Streets, 1877

Female, 7 mm; male, 18 mm

1   Lateral view, male
2   Antennula in lateral and internal view, with detail
3   Pereiopod 1, male, with detail
4   Pereiopod 2, male

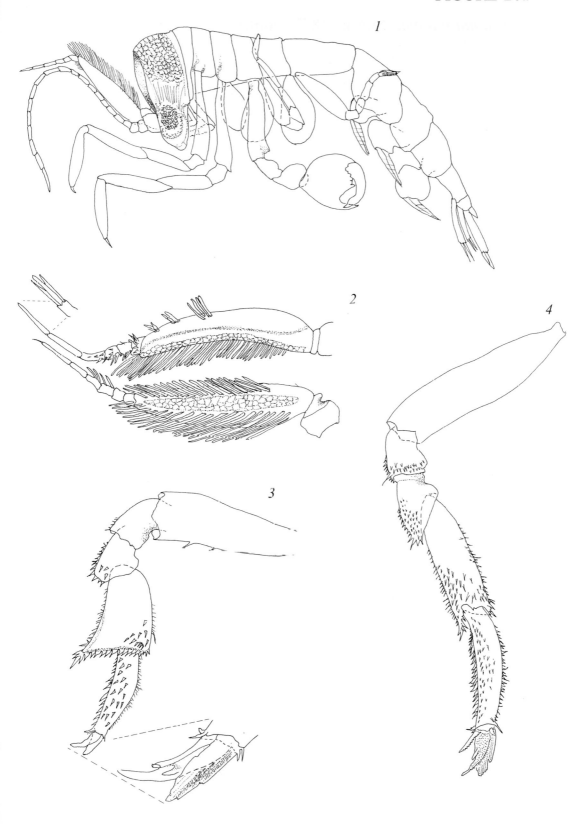

1

2

3

4

FIGURE 26b

Phronima pacifica Streets, 1877 (cont.)

5 Pereiopod 4, male, with detail
6 Pereiopod 5, male, with details of subchela
7 Pereiopod 6, male, with detail

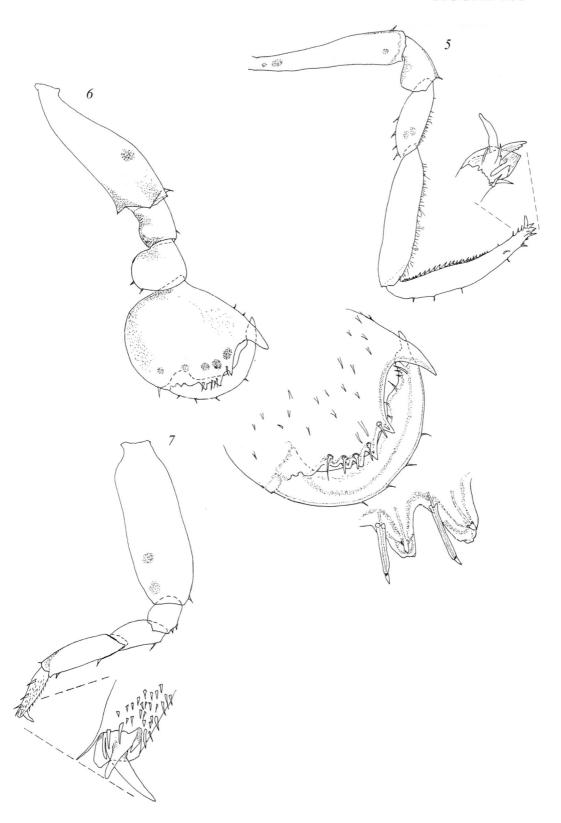

FIGURE 26c

Phronima pacifica Streets, 1877 (cont.)

8   Pereiopod 7, male, with detail
9   Pereiopod 1, female, dactylus
10 Pereiopod 2, female, dactylus
11 Pereiopod 3, female, dactylus
12 Pereiopod 4, female, dactylus
13 Uropod 2

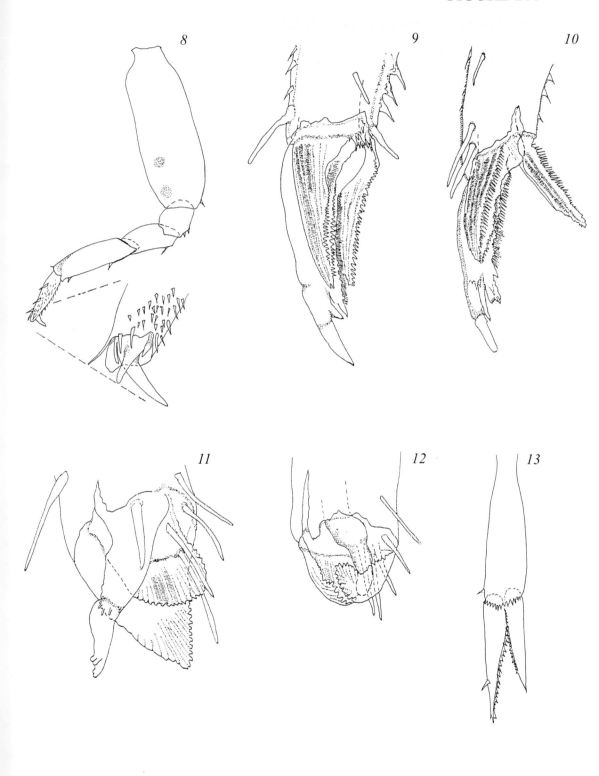

FIGURE 27a

Phronimella elongata (Claus, 1862)

Females, 7 and 12 mm

1 Lateral view
2 Antennula
3 Mandibula
4 Maxillula, 7 mm specimen, flattened
5 Maxillular palp, 12 mm specimen
6 Maxilla and maxilliped in anterior view

FIGURE 27a

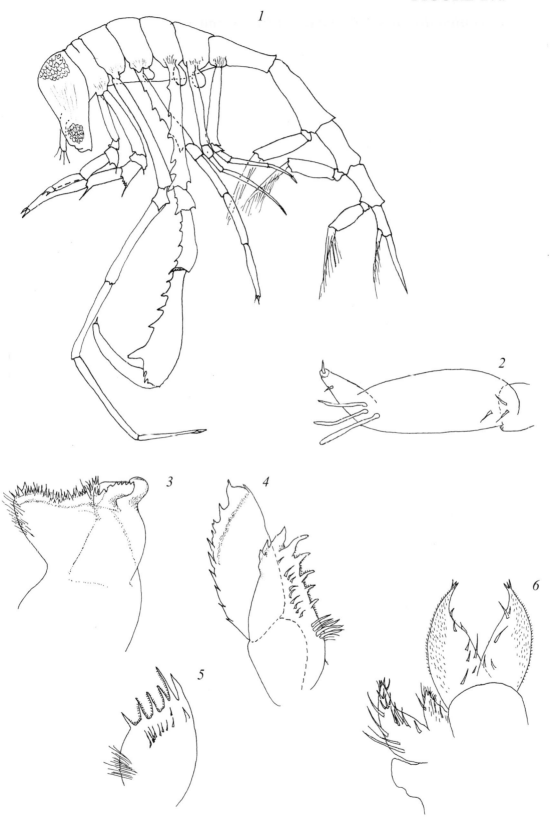

FIGURE 27b

Phronimella elongata (Claus, 1862) (cont.)

7a Pereiopod 1, 7 mm specimen, with detail
7b Fragment of propus and dactyl, another view
8   Pereiopod 1, only enlarged propus and dactylus
9   Pereiopod 2, with detail
10 Pereiopod 2, 12 mm specimen, enlarged propus and dactylus
11 Pereiopod 3

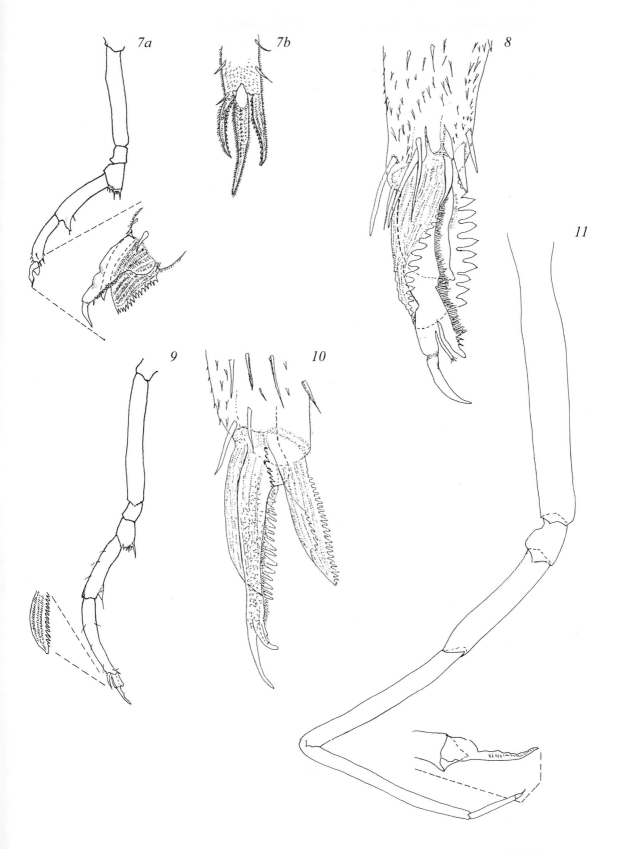

FIGURE 27c

Phronimella elongata (Claus, 1862) (cont.)

12 Pereiopod 4, with detail
13 Pereiopod 5, proximal part
14 Pereiopod 5, distal part
15 Pereiopod 7, with details of left and right dactylus
16 Uropods and telson, with detail

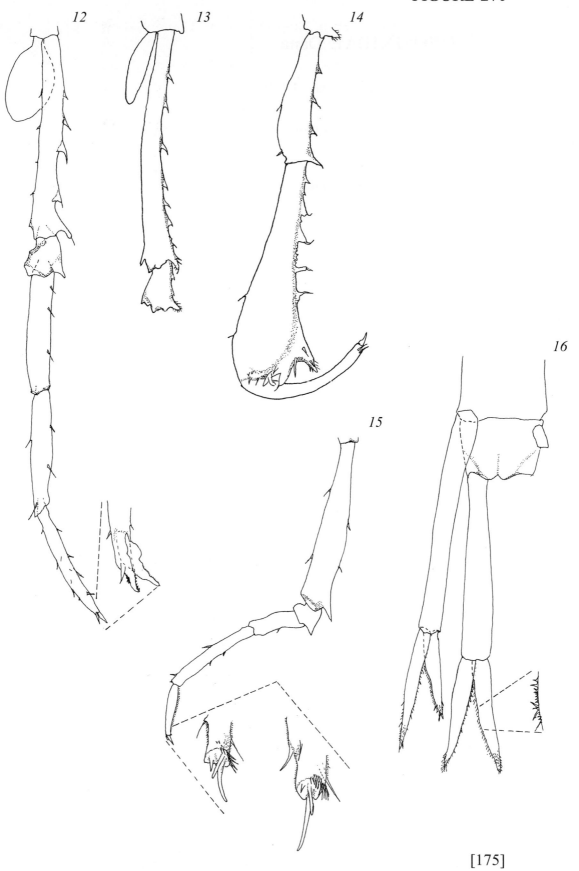

FIGURE 28a

PHROSINIDAE Dana

Phrosina semilunata Risso, 1822

Females, 4 mm and 12 mm; male, 6 mm

1   Female, lateral view
2   Female, 4 mm, dorsal view
3   Male, head
4   Mandibula

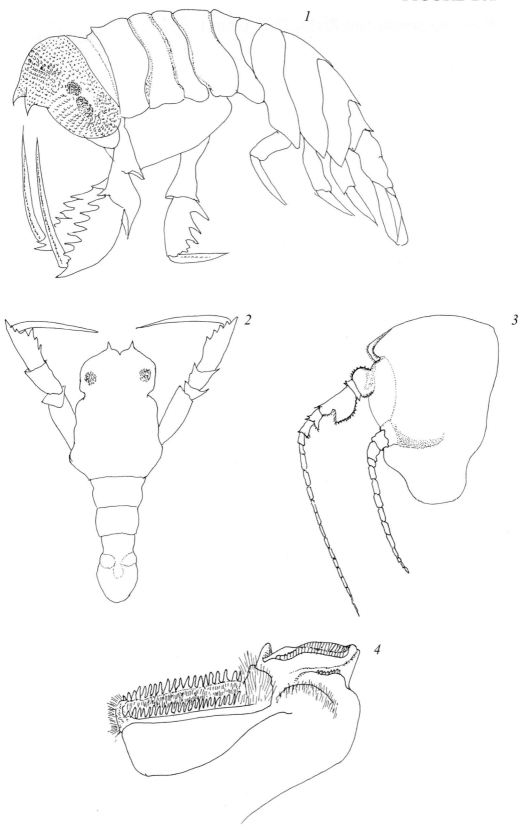

FIGURE 28b

Phrosina semilunata Risso, 1822 (cont.)

5   Maxillula
6   Maxilla
7   Pereiopod 1, female, with details
8   Pereiopod 2, female
9   Pereiopod 3, female

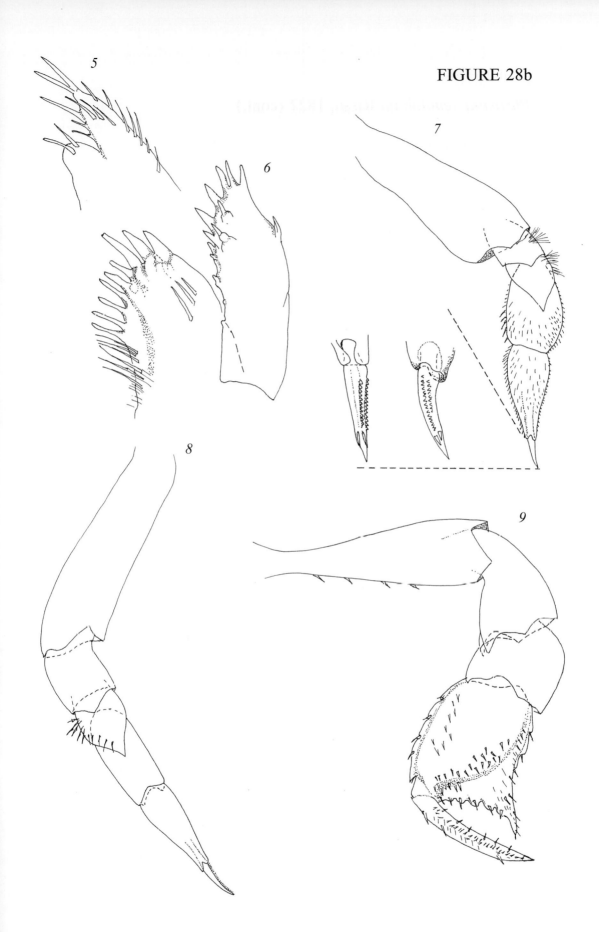

FIGURE 28b

FIGURE 28c

Phrosina semilunata Risso, 1822 (cont.)

10 Pereiopod 4, female
11 Pereiopod 4, male
12 Pereiopod 5, female
13 Pereiopod 5, male
14 Pereiopod 6, female
15 Pereiopod 7, female

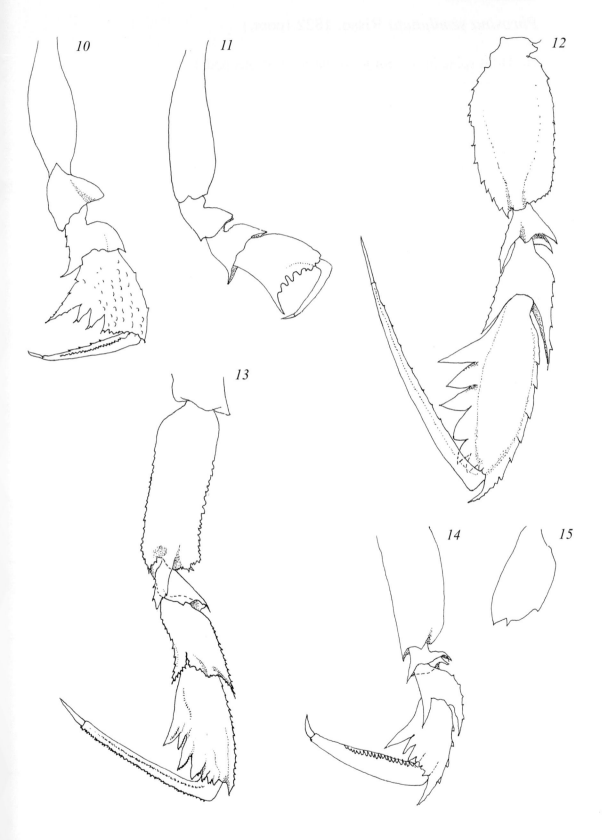

FIGURE 28d

Phrosina semilunata Risso, 1822 (cont.)

16 Bifid spine in the border of basipod of pleopod 1
17 Coupling hooks
18 Uropods and telson, female

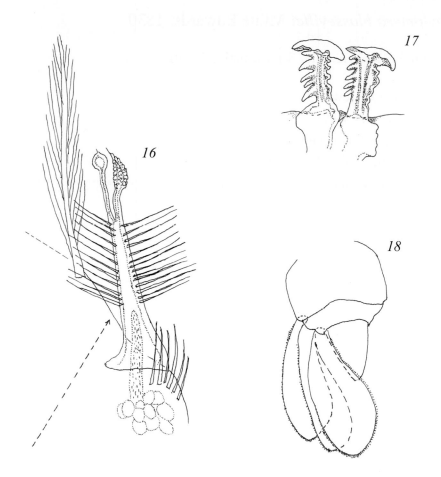

FIGURE 29a

Anchylomera blossevillei Milne Edwards, 1830

Specimen 1.5 mm; female, 8 mm; male, 9 mm

1   Male, lateral view
2   Female, dorsal view
3   1.5 mm specimen, lateral view
4   1.5 mm specimen, dorsal view
5   Antennula, female
6   Mandibula, female, external and internal view

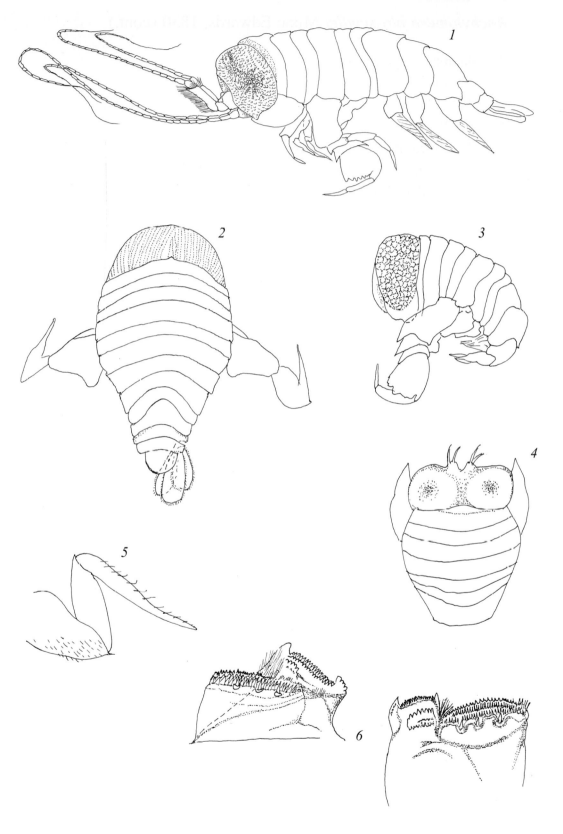

FIGURE 29b

Anchylomera blossevillei Milne Edwards, 1830 (cont.)

7   Mandibula, male
8   Maxillula, female
9   Maxillula, male, with detail
10 Maxilla, male
11 Maxilliped, male, lateral view
12 Maxilliped, female, lateral view
13 Pereiopod 1, female
14 Pereiopod 2, female

FIGURE 29b

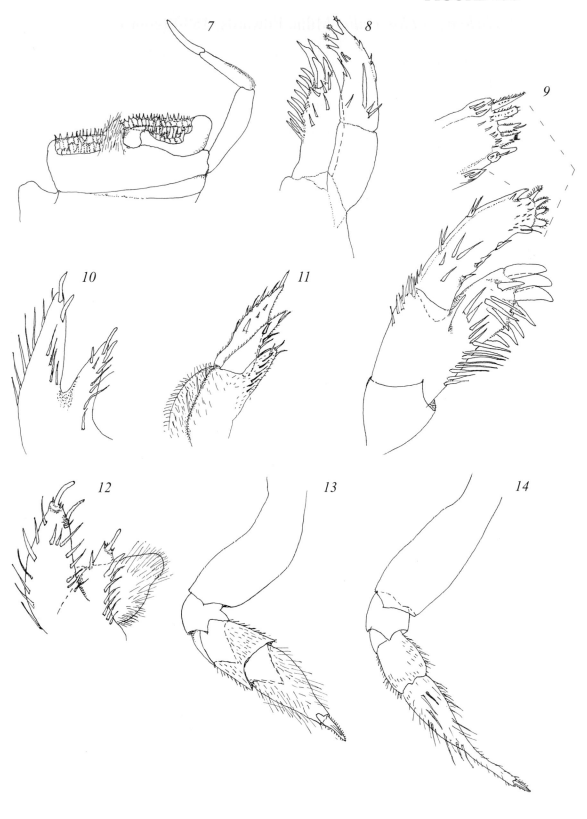

FIGURE 29c

Anchylomera blossevillei Milne Edwards, 1830 (cont.)

15 Pereiopod 3, female
16 Pereiopod 4, female
17 Pereiopod 4, male, with detail
18 Pereiopod 5, female, with detail

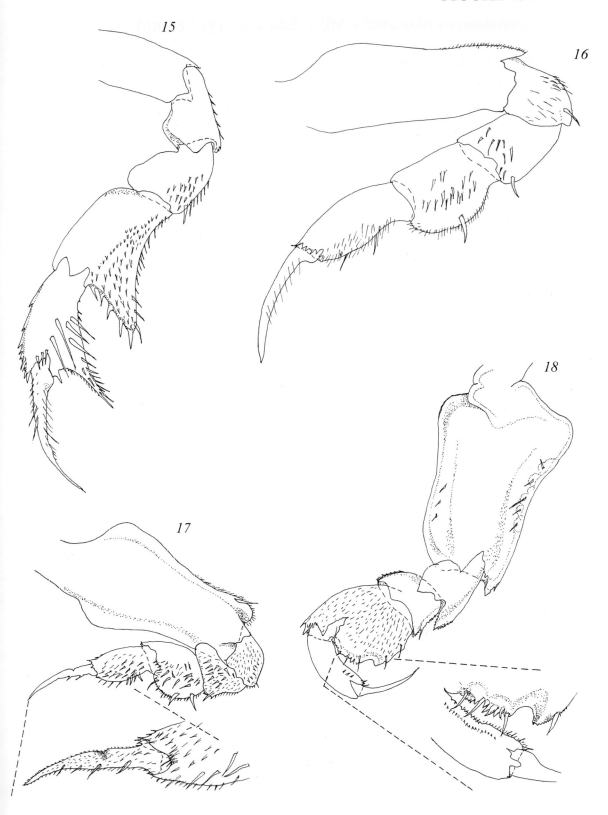

FIGURE 29d

Anchylomera blossevillei Milne Edwards, 1830 (cont.)

19 Pereiopod 5, male
20 Pereiopod 6, female
21 Pereiopod 6, male
22 Pereiopod 7, female, with detail
23 Pereiopod 7, male
24 Uropods and telson

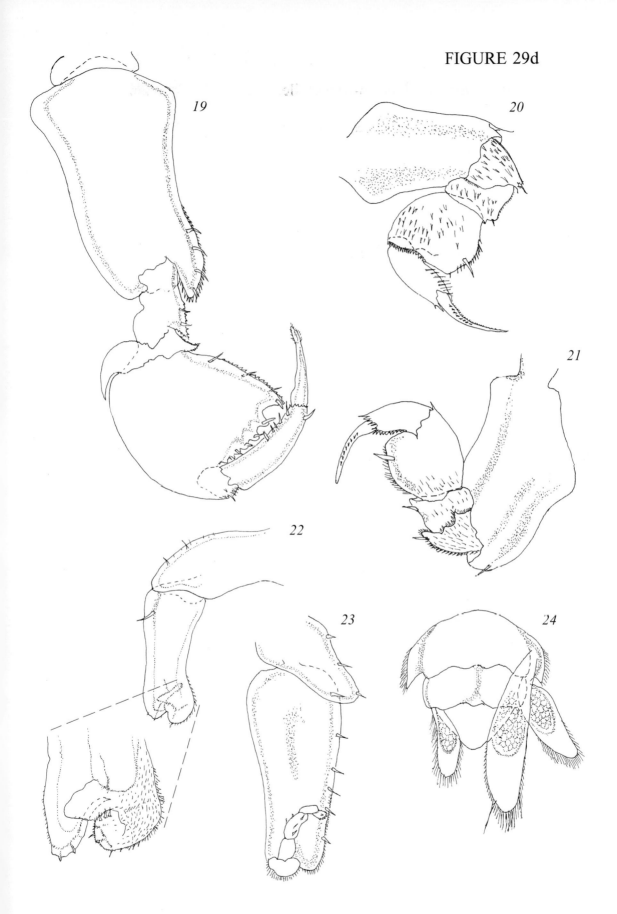

FIGURE 30a

Primno macropa Guérin-Ménéville, 1836

Females, 2 and 8 mm; males, 4 and 7 mm

1   Female, 8 mm, lateral view
2   Male, 7 mm, lateral view
3   Female, 8 mm, dorsal view
4   Head, male 7 mm
5   Antennula, female 8 mm

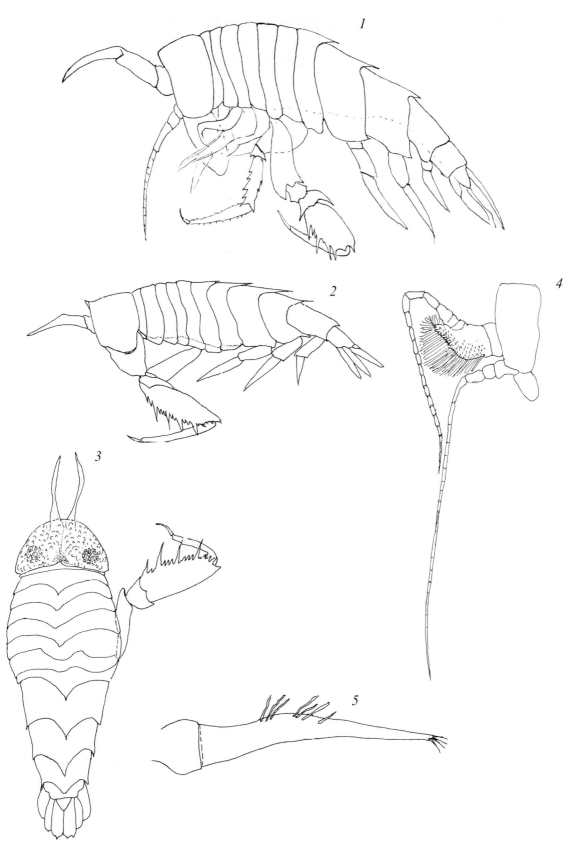

FIGURE 30a

FIGURE 30b

Primno macropa Guérin-Ménéville, 1836 (cont.)

6   Mandibulae, left and right, female 2 mm
7   Maxillulae, left and right, female 2 mm
8   Maxilla, female 2 mm
9   Maxilliped, female 2 mm
10 Pereiopod 1, female 8 mm, with detail
11 Pereiopod 1, female 2 mm, with detail
12 Pereiopod 2, female 2 mm

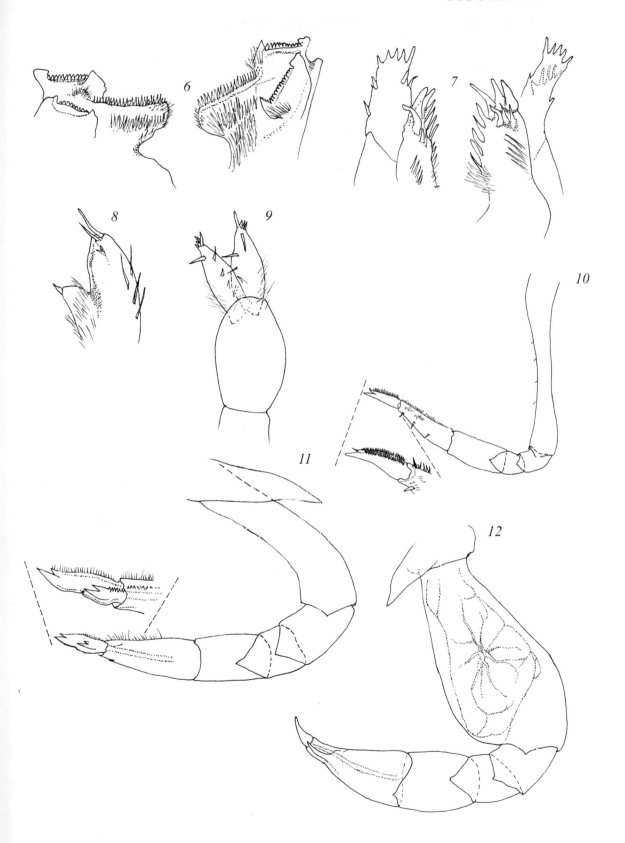

FIGURE 30c

Primno macropa Guérin-Ménéville, 1836 (cont.)

13 Pereiopod 5, female 8 mm, with detail
14 Pereiopod 5, male 7 mm, with detail
15 Pereiopod 6, female 8 mm
16 Pereiopod 7, female 8 mm, with detail

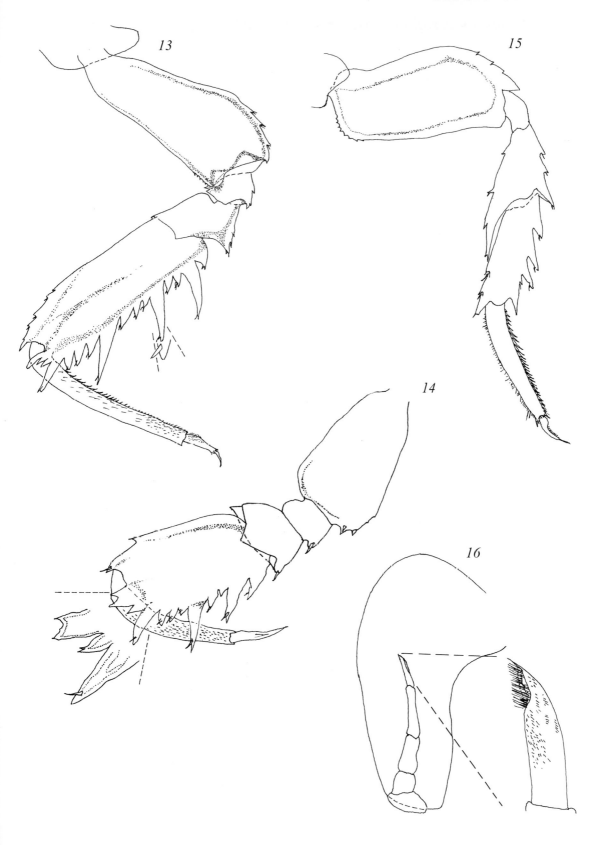

FIGURE 30d

Primno macropa Guérin-Ménéville, 1836 (cont.)

17 Pereiopod 7, male
18 Epimere 3
19 Pleopod 1, first segment with aesthetascs (?)
    (in depression)
20 Pleopod 3, female 8 mm, with bifid spine
    and coupling hooks
21 Uropods and telson

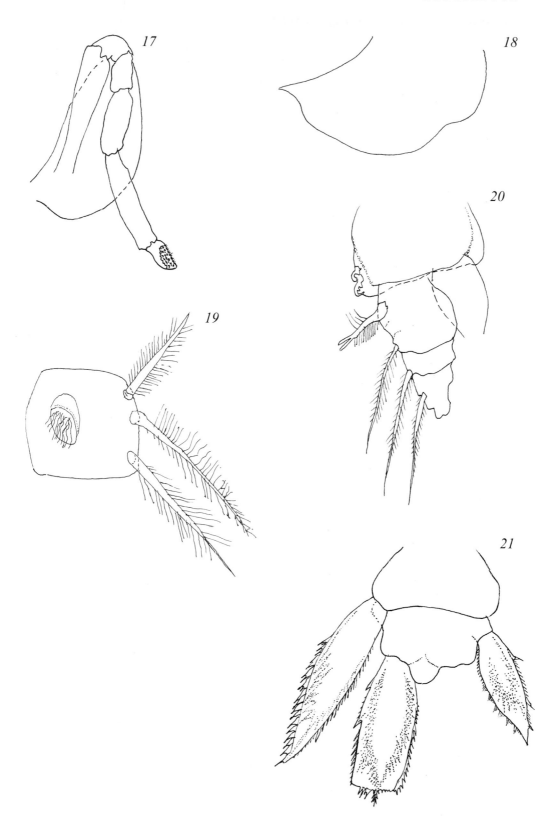

FIGURE 31a

Primno brevidens Bowman, 1978

Females, 3.5 and 5 mm

1   Mandibula, female 5 mm
2   Maxillulae, left and right, female 5 mm
3   Maxilla, female 5 mm
4   Maxilliped, female 5 mm
5   Pereiopod 1, female 3.5 mm, with detail
6   Pereiopod 2, female 3.5 mm
7   Pereiopod 3, female 3.5 mm

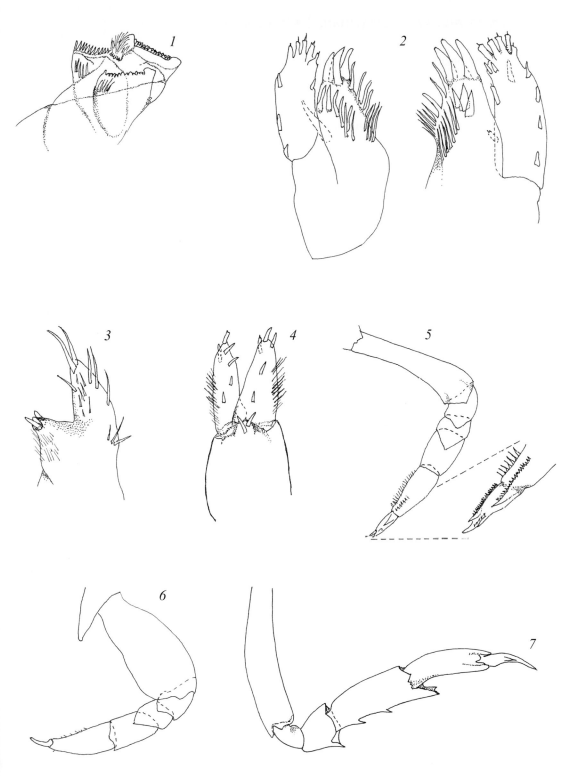

FIGURE 31b

Primno brevidens Bowman, 1978 (cont.)

8   Pereiopod 5, female 3.5 mm, with detail
9   Pereiopod 6, female 3.5 mm
10 Pereiopod 7, female 3.5 mm
11 Uropods and telson, with details

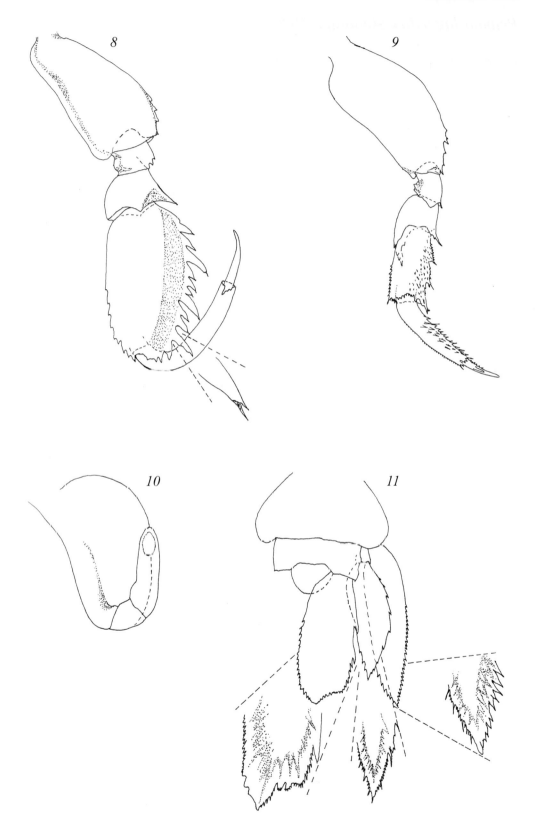

FIGURE 32a

Primno lattreillei Stebbing, 1888

Females, 7.5 and 8.5 mm; males, 3 and 8 mm

1   Head, male 3 mm
2   Antennula, female 7.5 mm
3   Antennula, male
4   Antenna, male 8 mm
5   Mandibula, female 7.5 mm
6   Maxillula, female 7.5 mm
7   Maxilla, female 7.5 mm
8   Maxilliped, female 7.5.mm

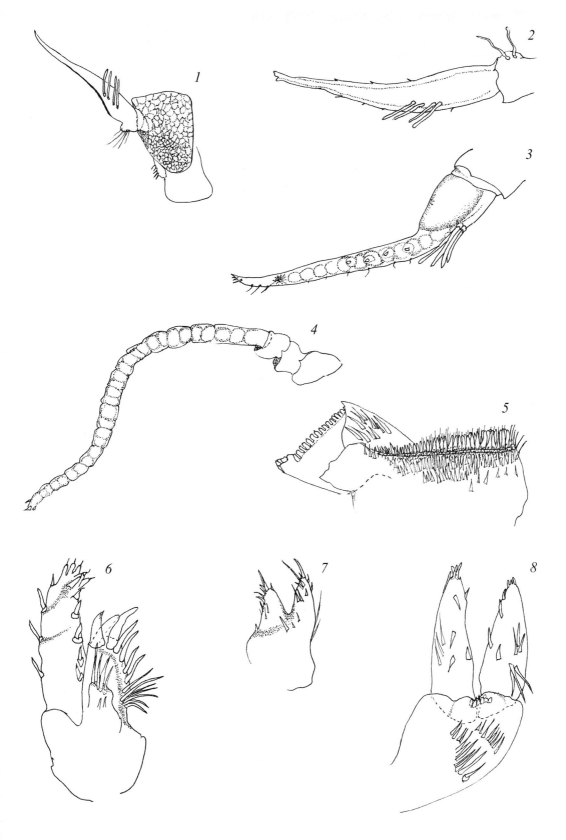

FIGURE 32b

Primno lattreillei Stebbing, 1888 (cont.)

9 Pereiopod 1, female 7.5 mm, with detail
10 Pereiopod 2, female 7.5 mm
11 Pereiopod 3, female 7.5 mm
12 Pereiopod 4, female 7.5 mm
13 Pereiopod 5, female 7.5 mm, with details

FIGURE 32b

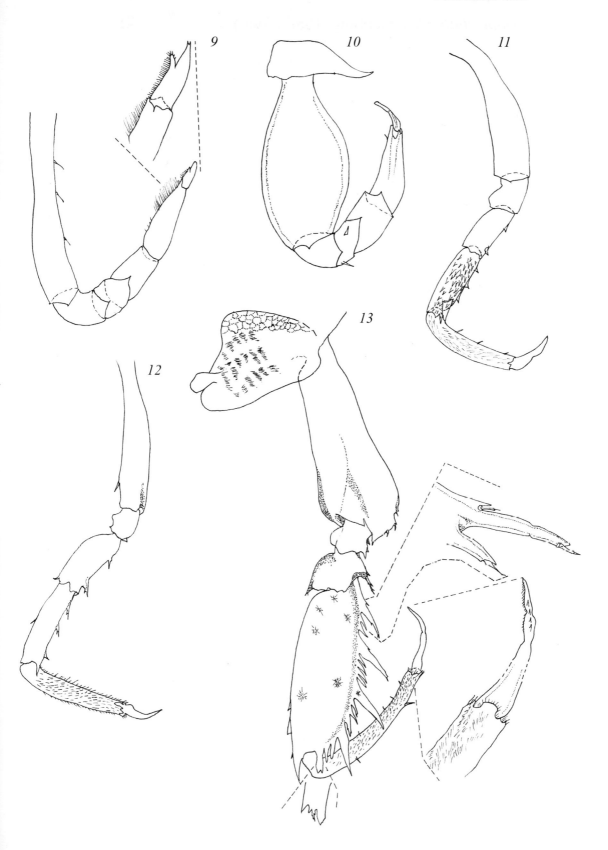

FIGURE 32c

Primno lattreillei Stebbing, 1888 (cont.)

14 Pereiopod 6, female 7.5 mm, with detail
15 Pereiopod 7, female 7.5 mm, with detail
16 Pereiopod 7, male 8 mm
17 Epipod of pereiopod 6, female
18 Epimere 3, posterolateral angle
19 Uropods and telson, female 7.5 mm
20 Serration details of uropod 1, uropod 2, uropod 3

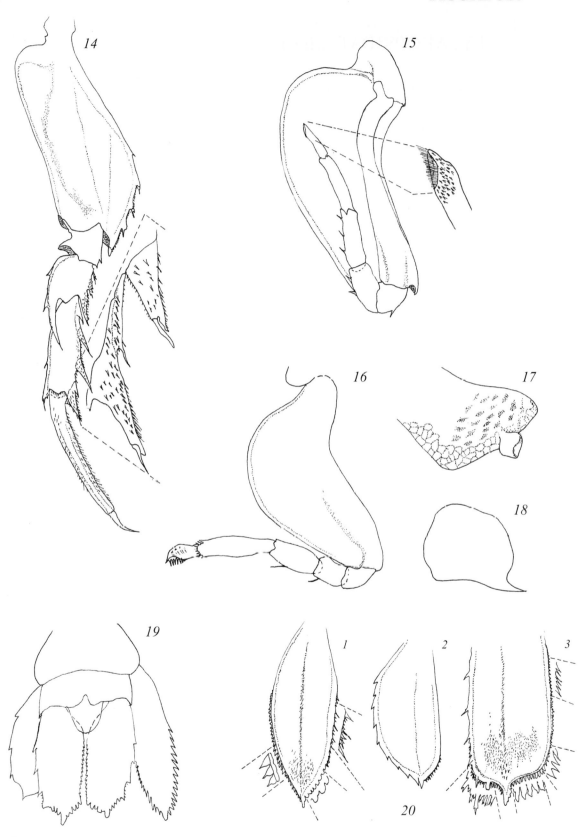

FIGURE 33a

LYCAEOPSIDAE Chevreux

Lycaeopsis themistoides Claus, 1879

Females, 3.5 and 4 mm; male 5 mm

1   Female, 3.5 mm, lateral view
2   Male, lateral view
3   Antennula, female
4   Antennula, male
5   Antenna, male
6   Maxilliped, female, with detail
7   Maxilliped, male

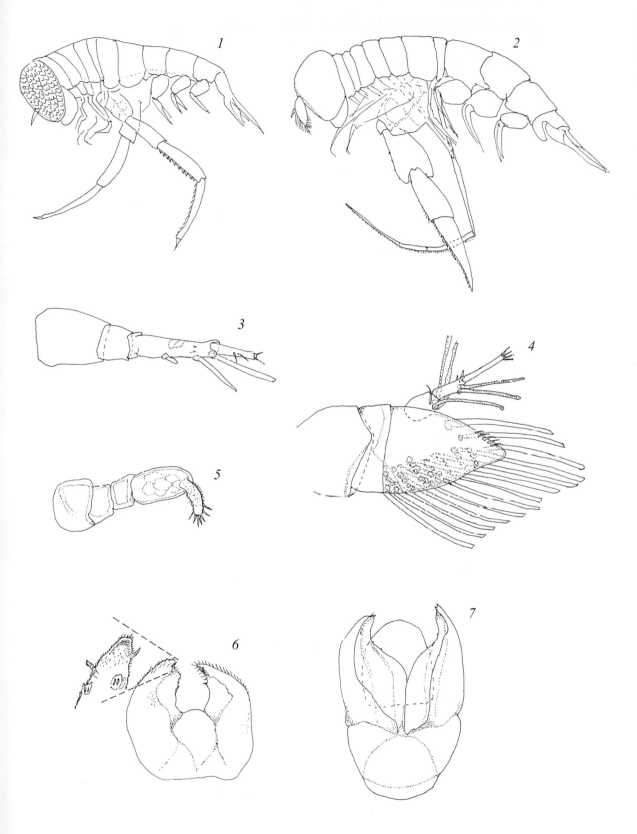

FIGURE 33b

Lycaeopsis themistoides Claus, 1879 (cont.)

8   Pereiopod 1, female, with detail
9   Pereiopod 1, male, with detail
10 Pereiopod 2, female, with detail
11 Pereiopod 2, male, with detail
12 Pereiopod 3, female, with detail
13 Pereiopod 3, male, with detail

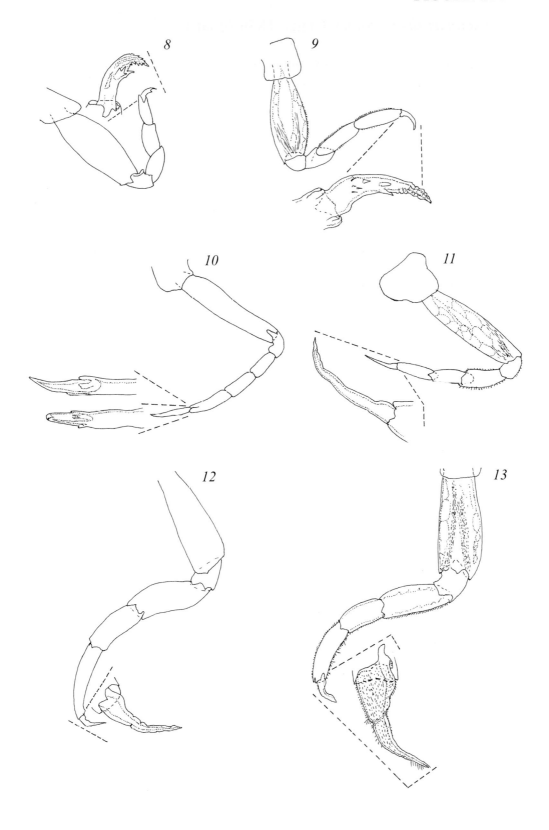

FIGURE 33c

Lycaeopsis themistoides Claus, 1879 (cont.)

14 Pereiopod 4, female, with detail
15 Pereiopod 4, male, with details
16 Pereiopod 5, female, with detail
17 Pereiopod 5, male

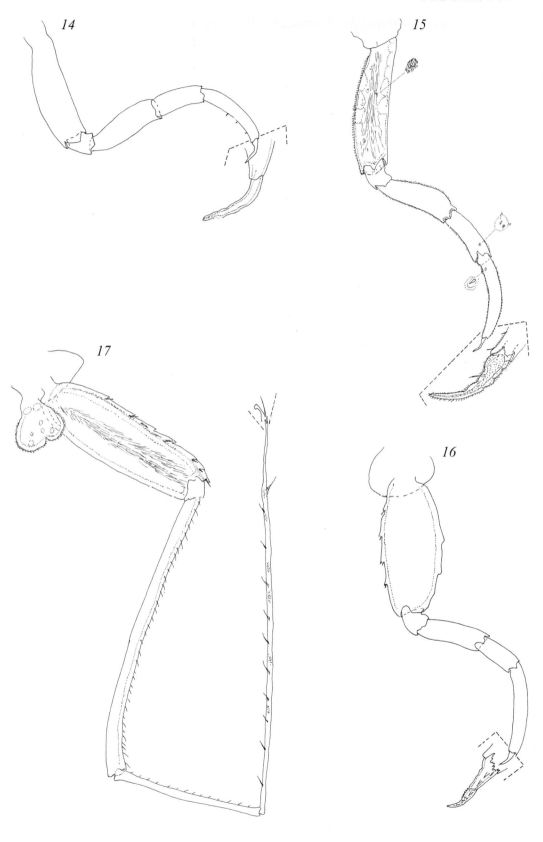

FIGURE 33c

14

15

17

16

FIGURE 33d

Lycaeopsis themistoides Claus, 1879 (cont.)

18 Pereiopod 6, female, with details
19 Pereiopod 6, male, with details
20 Pereiopod 7, female, with detail
21 Pereiopod 7, male, with detail

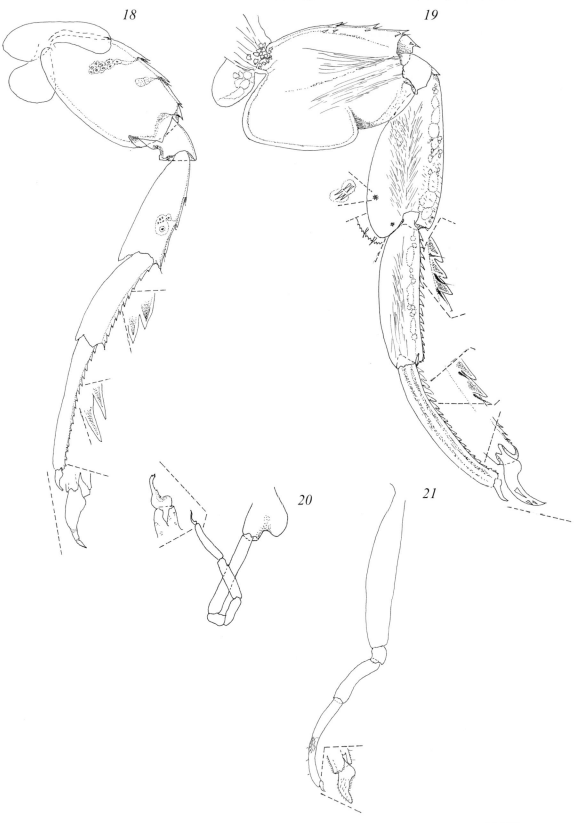

18

19

20

21

FIGURE 33e

Lycaeopsis themistoides Claus, 1879 (cont.)

22 Pleopod 1, proximal segments, with bifid spine and glandular structures
23 Uropods and telson, female, with details
24 Uropods and telson, male, with detail
25 Bifid spine of pleopod 1, male
26 Bifid spine of pleopod 2, male

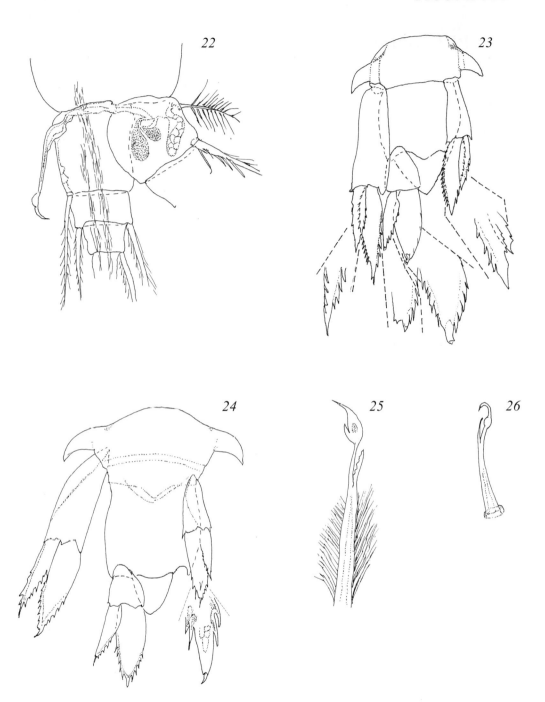

22

23

24

25

26

FIGURE 34

PRONOIDAE Claus

Eupronöe minuta Claus, 1879

Females, 4 mm; figures relatively enlarged !

1   Pereiopod 1, with details
2   Pereiopod 2, with details
3   Pereiopod 6, with details
4   Pereiopod 7

FIGURE 34

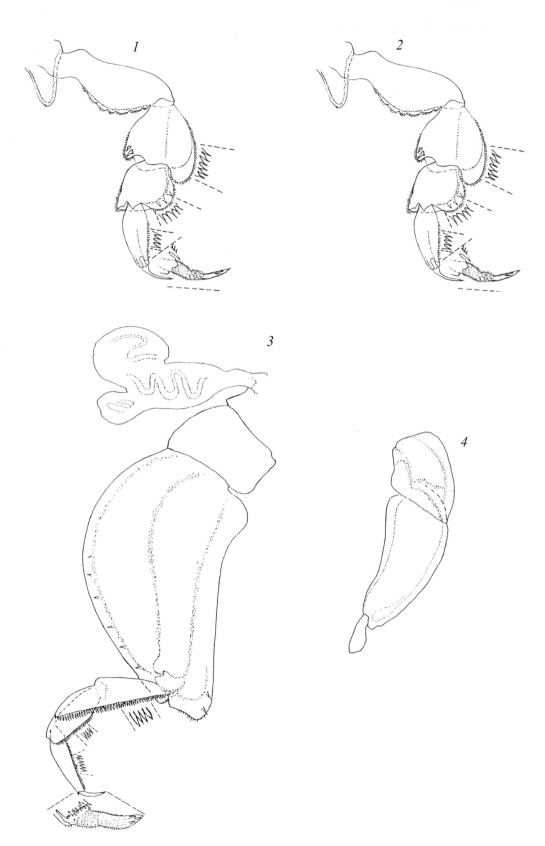

FIGURE 35a

Eupronöe armata Claus, 1879

Female, 7 mm

1   Lateral view (after Claus 1879)
2   Antennula
3   Mandibula and separately mandibular palp
4   Maxilliped (fragment: right outer lobe and inner lobe)
5   Pereiopod 1
6   Pereiopod 2, with detail
7   Pereiopod 3

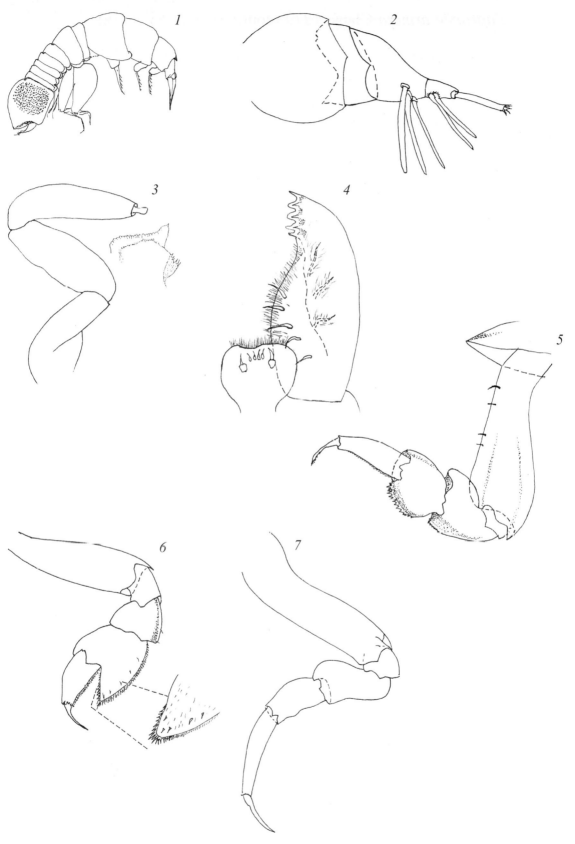

FIGURE 35b

Eupronöe armata Claus, 1879 (cont.)

8   Pereiopod 4
9   Pereiopod 5, with details
10   Pereiopod 6, with details
11a Pereiopod 7, left
11b Pereiopod 7, right
12   Uropods and telson
13   Uropod 1 and detail

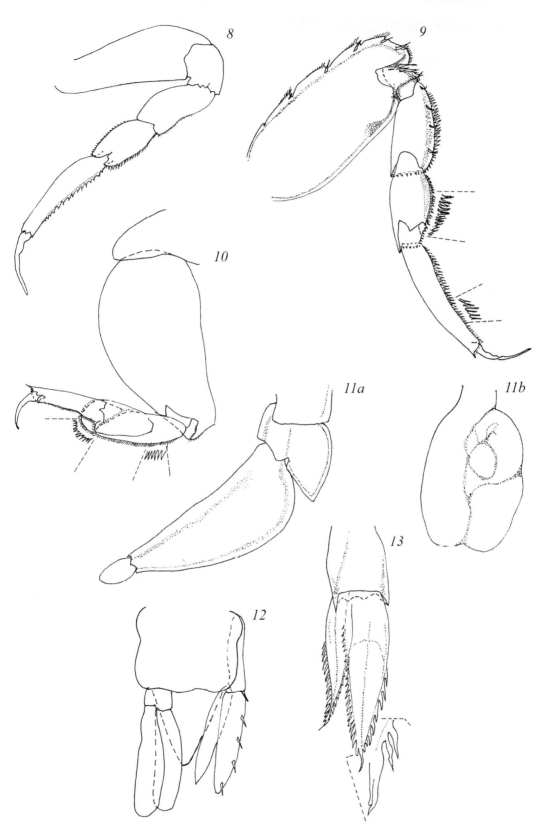

FIGURE 36a

Parapronöe parva Claus, 1879

Female, 6 mm

1   Mandibula
2   Maxillular palp
3   Maxilliped
4   Maxilliped, inner lobe
5   Pereiopod 1, with detail
6   Pereiopod 2, with details
7   Pereiopod 5

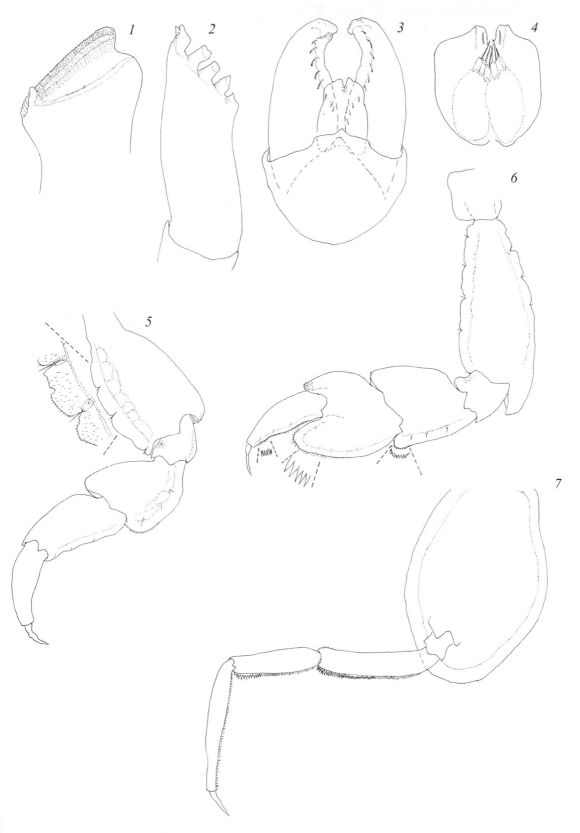

FIGURE 36b

Parapronöe parva Claus, 1879 (cont.)

8   Pereiopod 6, with details
9   Pereiopod 7
10 Uropods and telson

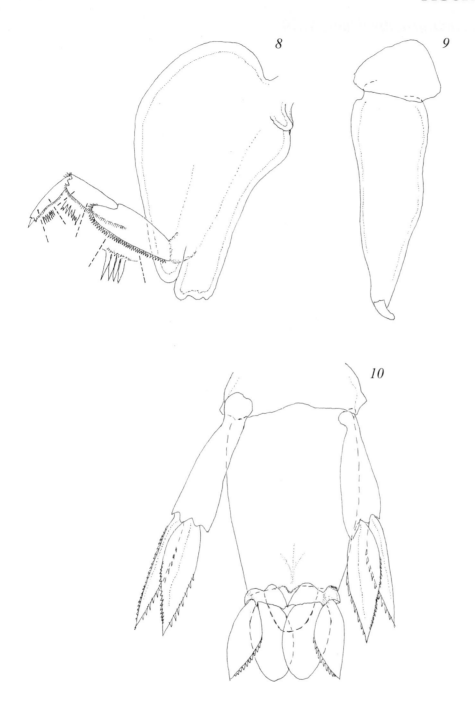

FIGURE 37a

Paralycaea gracilis Claus, 1879

Male, 5 mm

1   Lateral view
2   Antennula
3   Antenna
4   Mandibula with palp
5   Maxilliped
6   Pereiopod 1

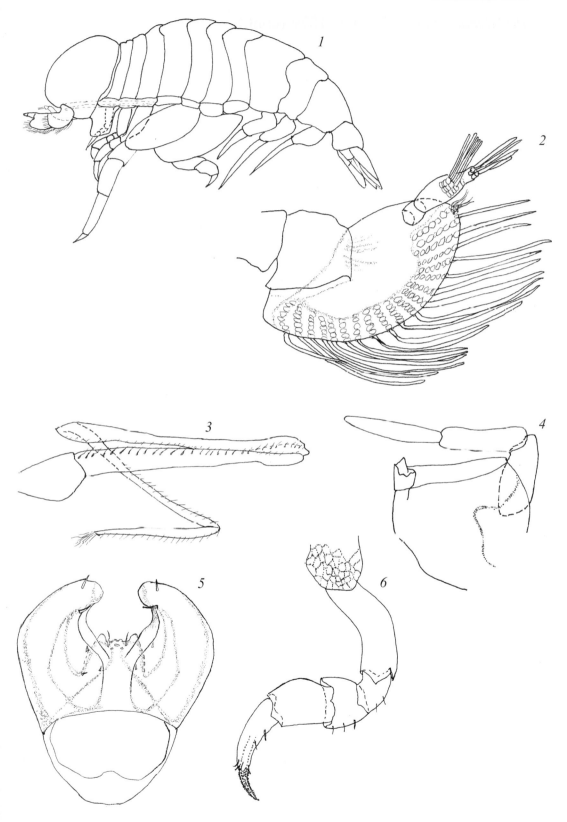

FIGURE 37a

FIGURE 37b

Paralycaea gracilis Claus, 1879 (cont.)

7   Pereiopod 2
8   Pereiopod 5
9   Pereiopod 6
10 Pereiopod 7
11 Uropods and telson

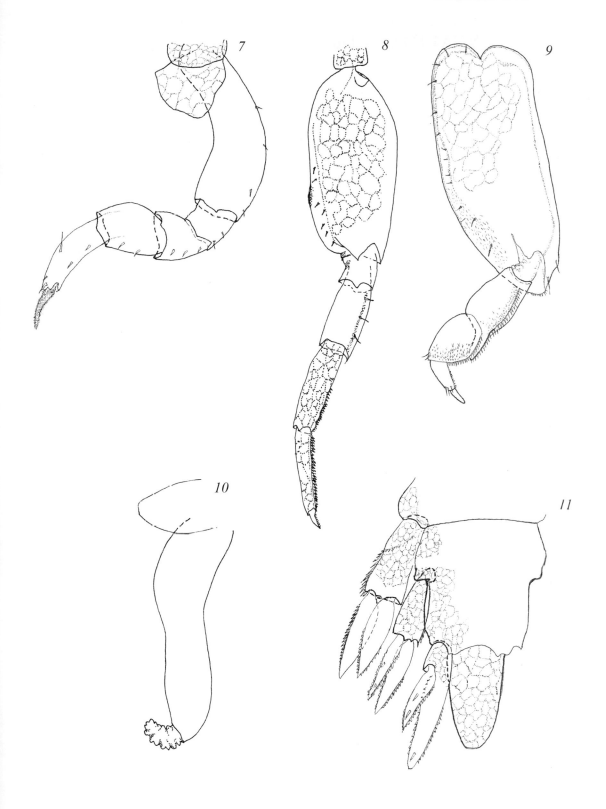

FIGURE 38a

LYCAEIDAE Claus

Lycaea pulex Marion, 1874

Female, 2 mm; male, 4 mm

1   Male, lateral view
2   Antennula, female
3   Antennula, male
4   Antenna, male
5   Mandibula, male, with palp
6   Maxilliped, female, posterior view

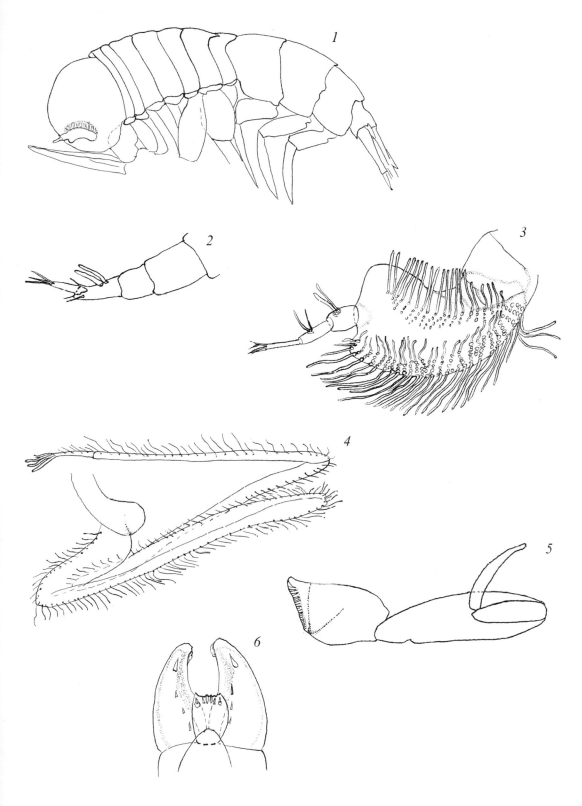

FIGURE 38b

Lycaea pulex Marion, 1874 (cont.)

7   Pereiopod 1, female

8   Pereiopod 1, male

9   Pereiopod 1, male, subchela with bifurcate and trifurcate spines on marginal carpal angle

10 Pereiopod 2, female

11 Pereiopod 2, male, with detail of spinulation

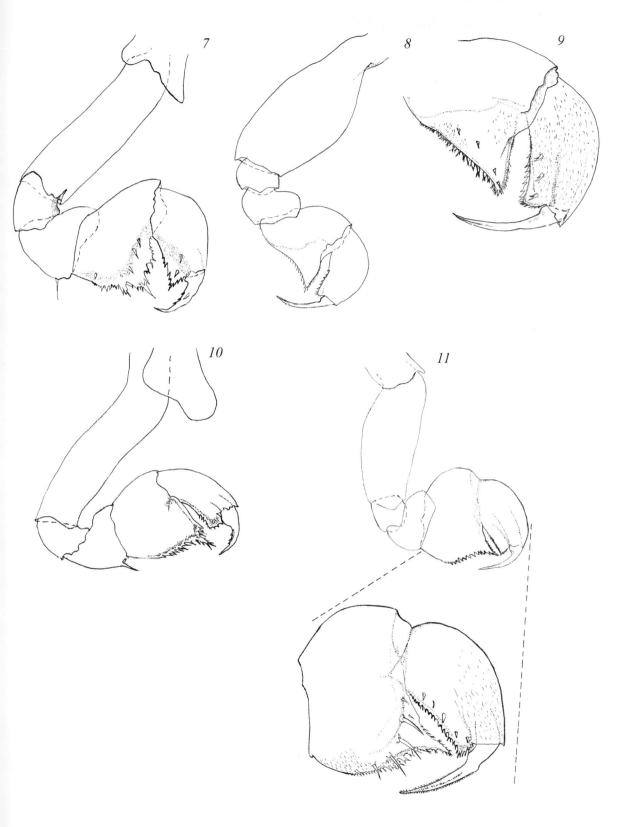

FIGURE 38c

Lycaea pulex Marion, 1874 (cont.)

12 Pereiopod 3, female
13 Pereiopod 4, female
14 Pereiopod 5, female
15 Pereiopod 5, male
16 Pereiopod 6, female, with details

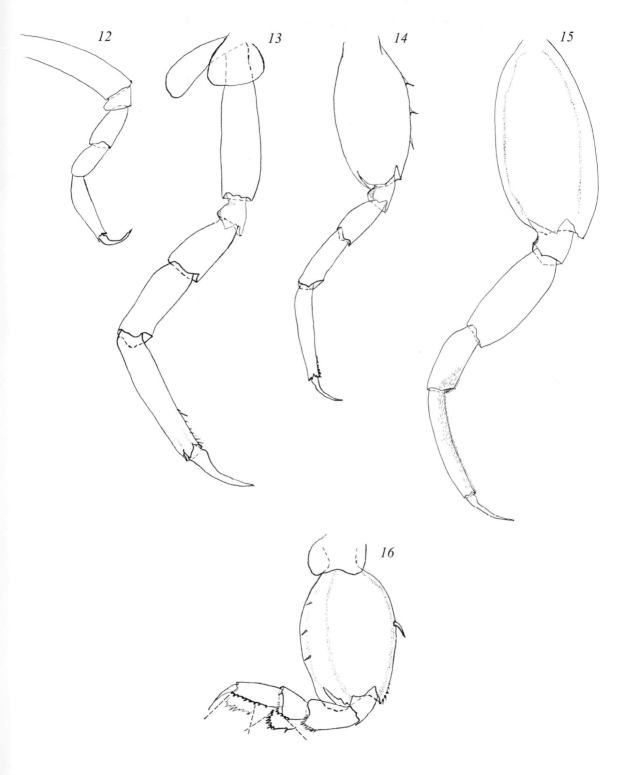

FIGURE 38d

Lycaea pulex Marion, 1874 (cont.)

17 Pereiopod 6, male, with detail
18 Pereiopod 7, female, with detail
19 Pereiopod 7, male, with detail
20 Uropods and telson, male

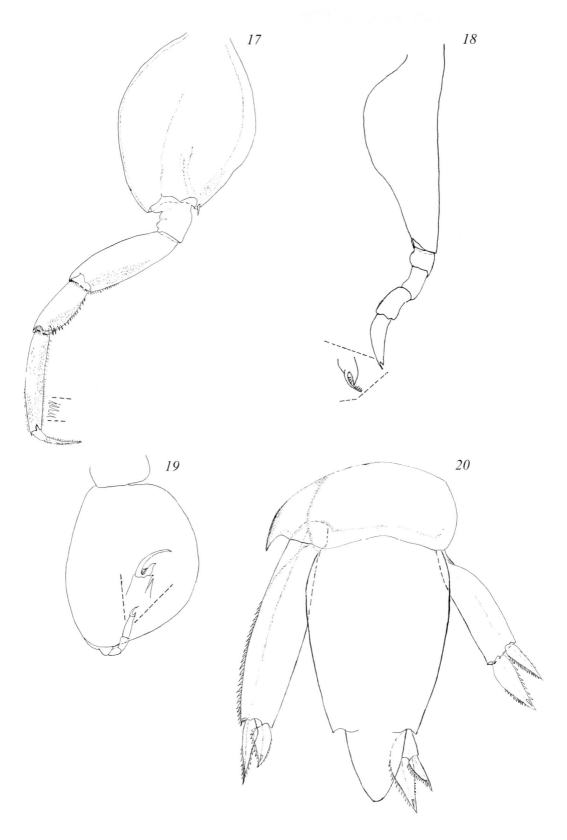

FIGURE 39a

Lycaea pauli Stebbing, 1888

Female, 4.5 mm; male, 6 mm

1   Female, lateral view
2   Male, lateral view
3   Antennula, female
4   Antennula, male
5   Antenna, male
6   Mandibula, female, gnathobase
7   Mandibula, male, with palp

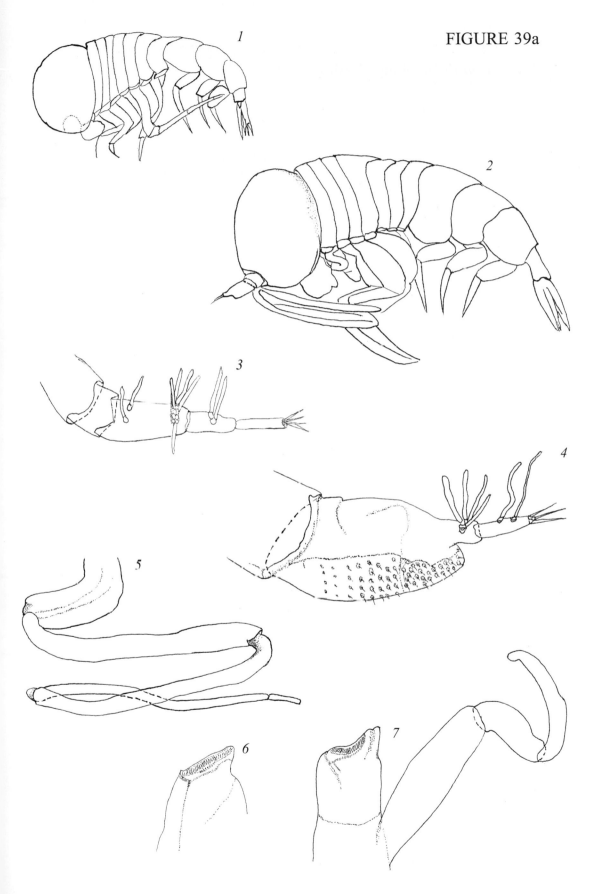

FIGURE 39a

FIGURE 39b

Lycaea pauli Stebbing, 1888 (cont.)

8   Maxilliped, female, anterior view
9   Maxilliped, male
10 Pereiopod 1, female
11 Pereiopod 1, male
12 Pereiopod 2, female, with detail

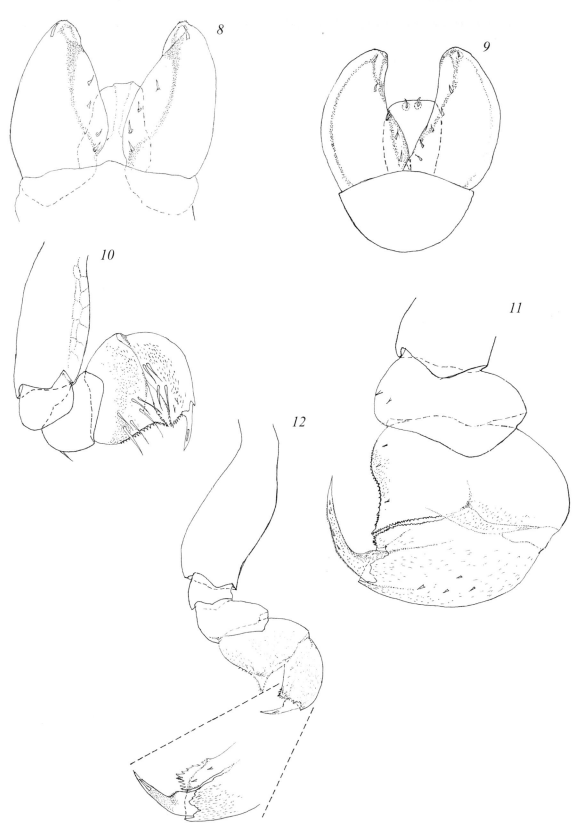

FIGURE 39c

Lycaea pauli Stebbing, 1888 (cont.)

13 Pereiopod 2, male, with epipod
14 Pereiopod 3, male
15 Pereiopod 4, male
16 Pereiopod 5, female, with detail

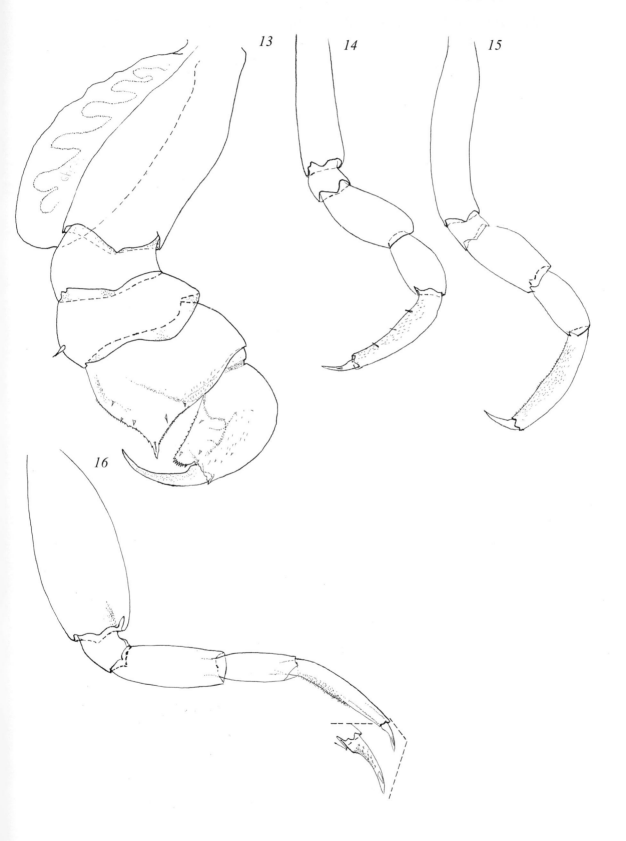

FIGURE 39d

Lycaea pauli Stebbing, 1888 (cont.)

17 Pereiopod 5, male
18 Pereiopod 6, female, with details
19 Pereiopod 6, male
20 Pereiopod 7, female, with detail
21 Pereiopod 7, male
22 Uropods and telson, male, with details

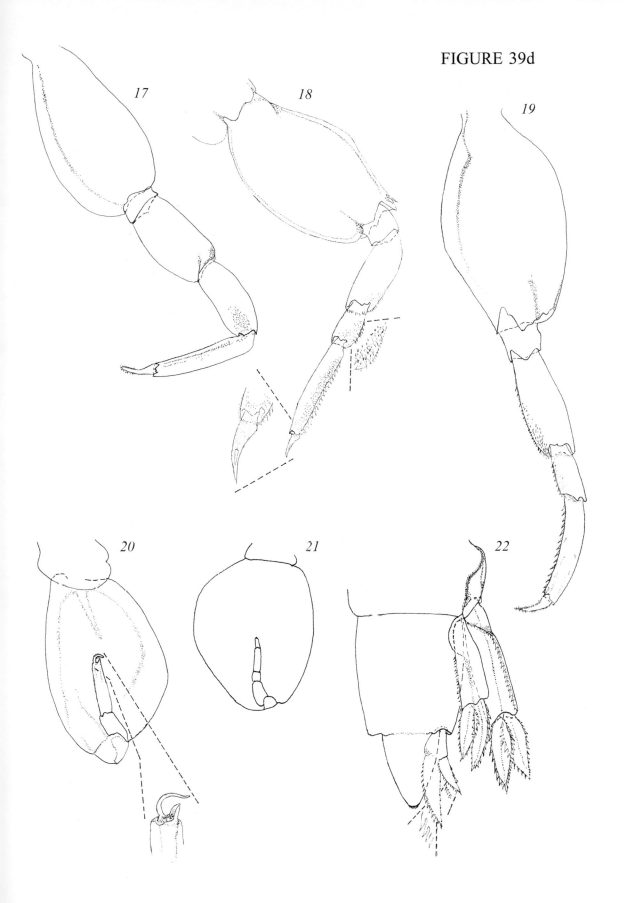

FIGURE 40a

Lycaea serrata Claus, 1879

Females, 7 and 8 mm

1   Lateral view
2   Antennula
3   Antennula, lower side
4   Mandibulae, left and right
5   Maxilliped
6   Maxilliped, fragment, lateral view
7   Pereiopod 1, with detail

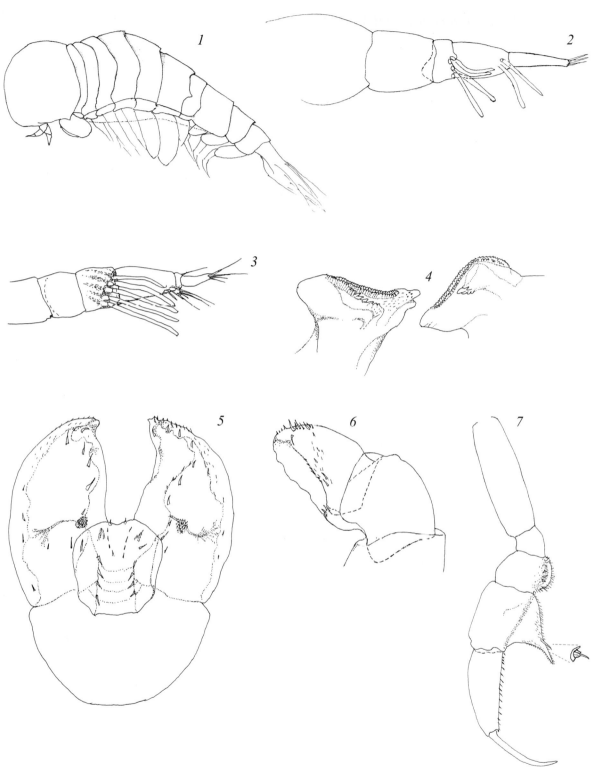

FIGURE 40a

[251]

FIGURE 40b

Lycaea serrata Claus, 1879 (cont.)

8   Pereiopod 2
9   Pereiopod 3
10 Pereiopod 4
11 Pereiopod 5
12 Pereiopod 6, with details
13 Pereiopod 7, with detail

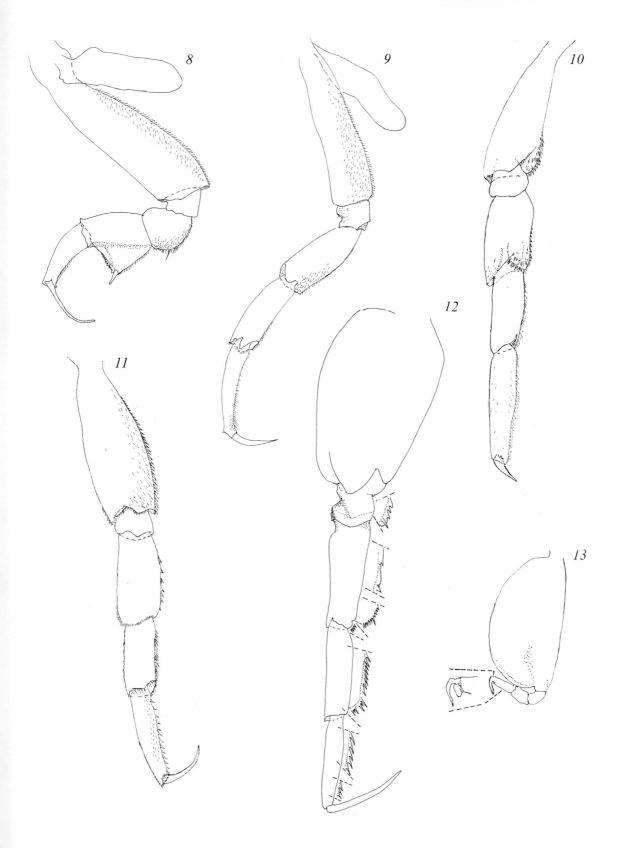

FIGURE 40c

Lycaea serrata Claus, 1879 (cont.)

14 Uropods and telson, with detail of spinulation on uropod 1
15 Bifid spines, triangular accessory spine and 3 coupling hooks on basipod of pleopod 1
16 The same on pleopod 2
17 The same on pleopod 3; on left side of exopod

Note: Nos. 15–17 can be seen only at high magnification.

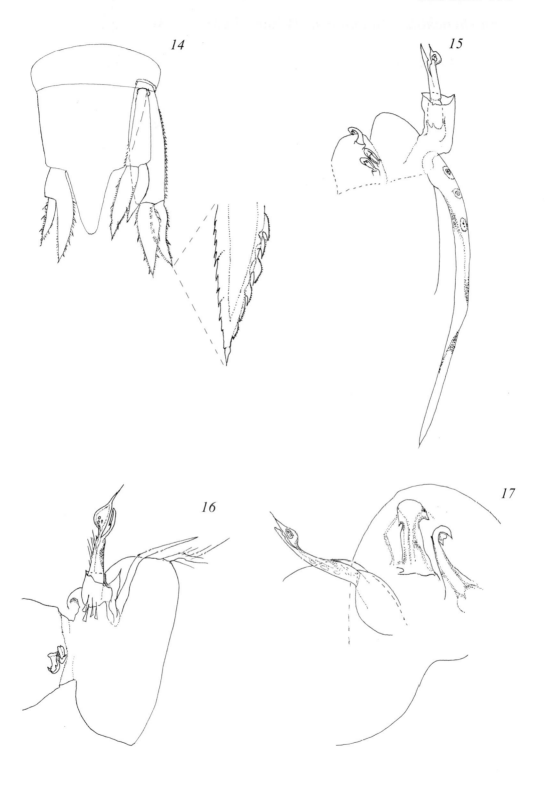

FIGURE 41a

Simorhynchotus antennarius (Claus, 1871)

Juvenile female, 2.5 mm; female, 7 mm; male, 4 mm

1   Female, lateral view
2   Antennula, female
3   Antenna, female
4   Mandibula, gnathobase, female
5   Mandibula, gnathobase, male
6   Mandibula, palp
7   Maxilliped, female
8   Pereiopod 1, female
9   Pereiopod 2, female

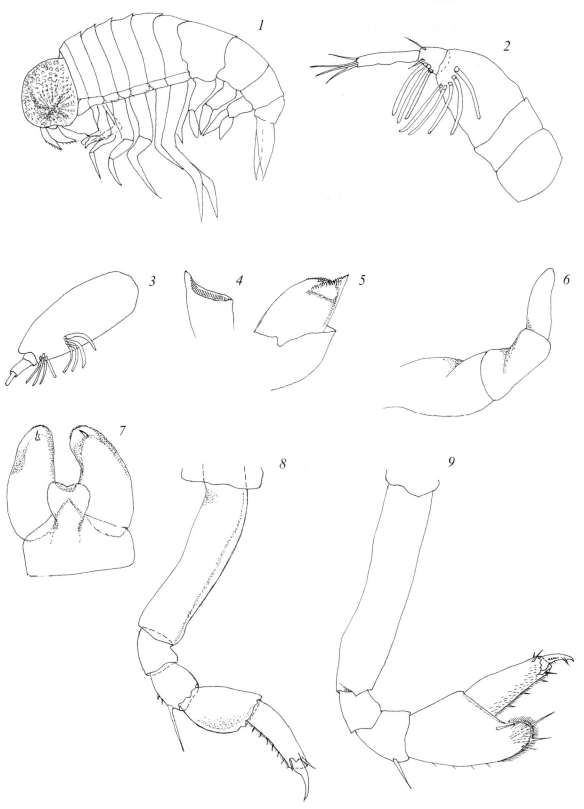

FIGURE 41b

Simorhynchotus antennarius (Claus, 1871) (cont.)

10 Pereiopod 4, female
11 Pereiopod 5, female, with detail
12 Pereiopod 6, juvenile female
13 Pereiopod 6, female
14 Pereiopod 7, juvenile female
15 Pereiopod 7, female
16 Uropods and telson, female

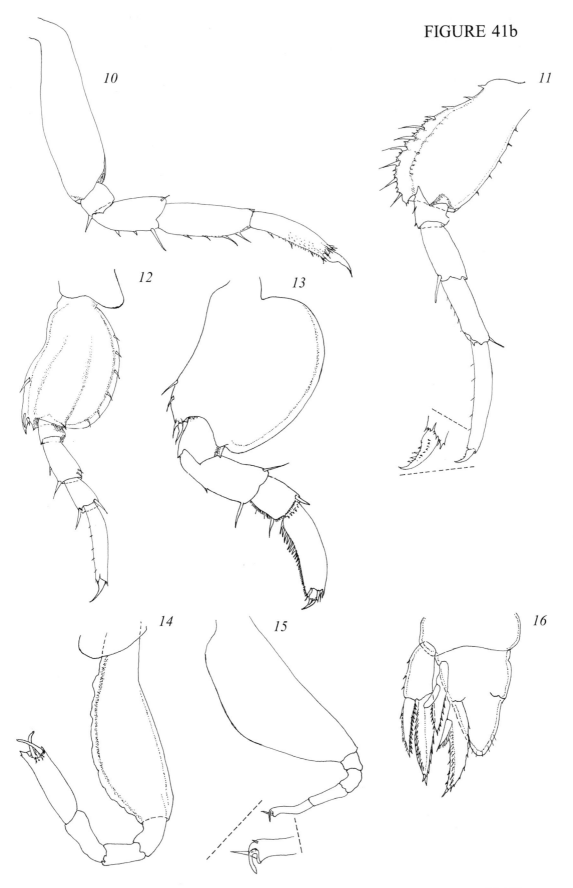

10

11

12

13

14

15

16

FIGURE 42a

BRACHYSCELIDAE Stephensen

Brachyscelus crusculum Bate, 1861

Females, 3.5 and 8 mm; males, 3.5 and 9 mm

1   Female 8 mm, lateral view
2   Antennula, female 8 mm
3   Antennula, male 9 mm, filaments partly omitted
4   Antenna, male 9 mm
5   Mandibular gnathobase, female
6   Mandibular gnathobase, male
7   Mandibular palp, male
8   Maxilliped, female 3.5 mm
9   Maxilliped, male 9 mm

FIGURE 42a

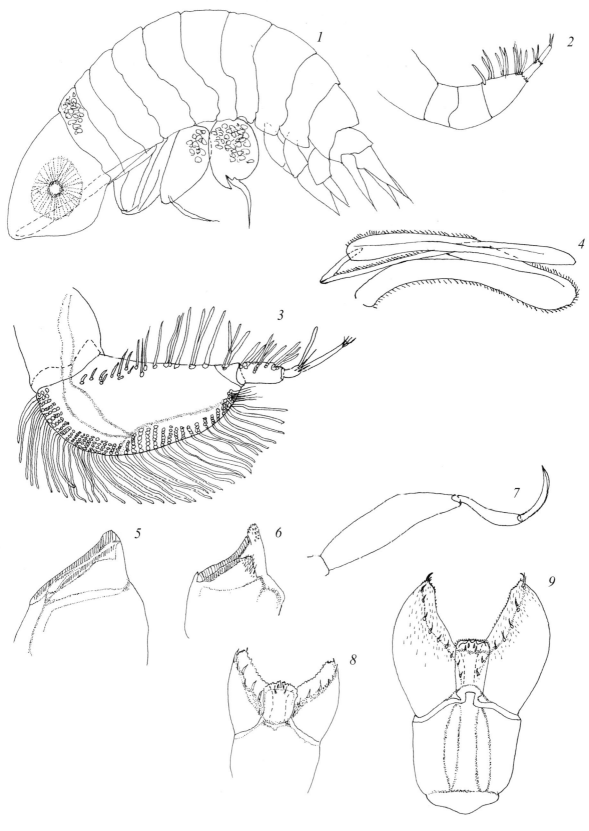

FIGURE 42b

Brachyscelus crusculum Bate, 1861 (cont.)

10 Pereiopod 1, male 3.5 mm, with detail
11 Pereiopod 1, male 9 mm, with details
12 Pereiopod 1, female 8 mm
13 Pereiopod 2, male 9 mm

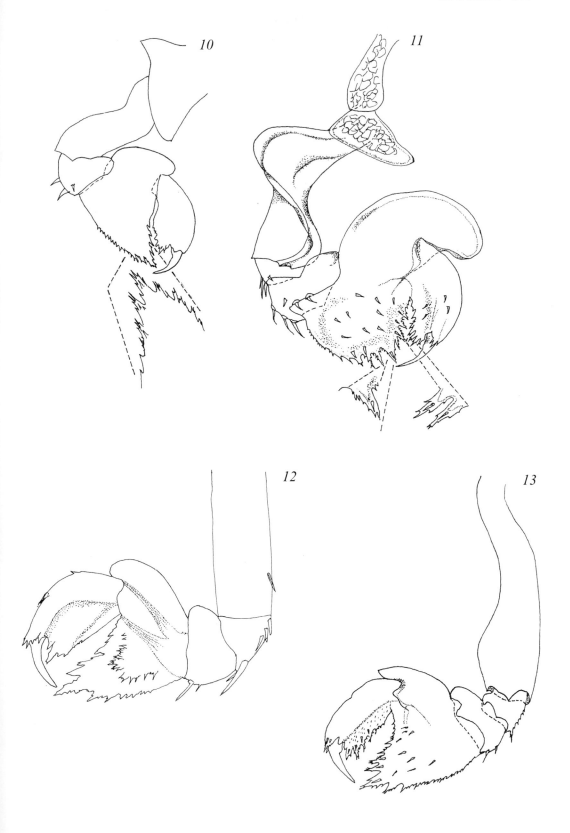

FIGURE 42c

Brachyscelus crusculum Bate, 1861 (cont.)

14 Pereiopod 2, female 8 mm
15 Pereiopod 3, female
16 Pereiopod 4, female
17 Pereiopod 4, male, with detail
18 Pereiopod 5, male, with epipodite and details

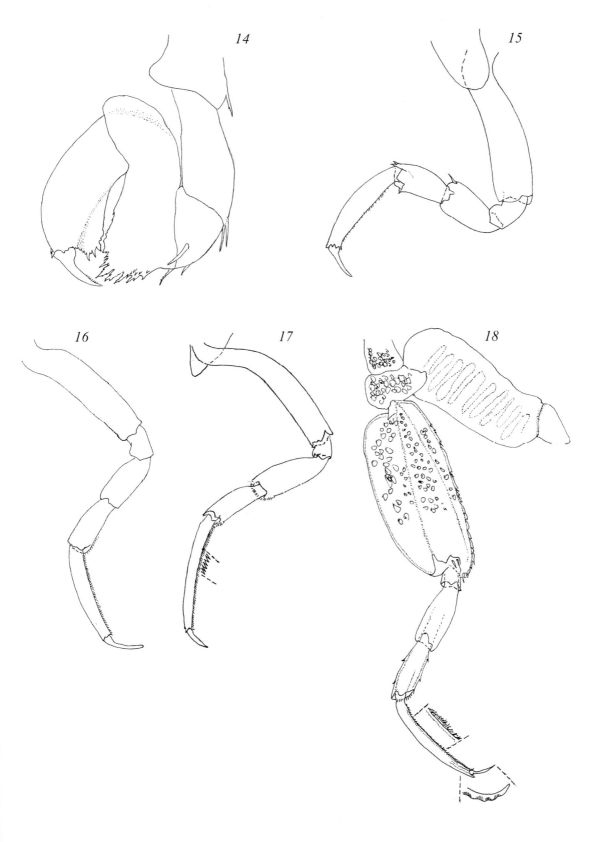

14

15

16

17

18

FIGURE 42d

Brachyscelus crusculum Bate, 1861 (cont.)

19 Pereiopod 5, female
20 Pereiopod 6, male, with epipodite and details
21 Pereiopod 6, female, with details
22 Pereiopod 7, male, with detail
23 Uropods and telson, male, with details
24 Uropod 3, male, enlarged with detail

FIGURE 42d

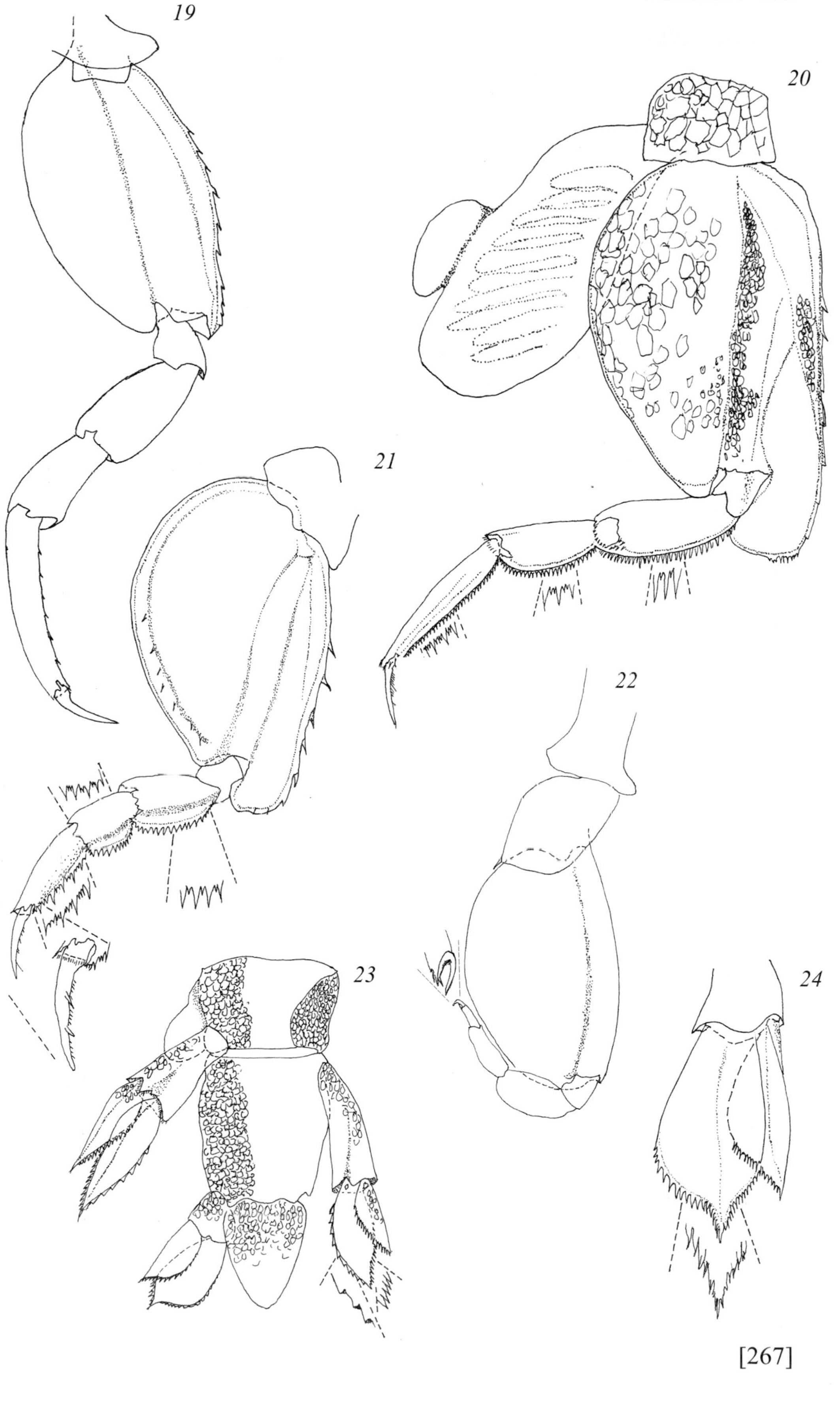

FIGURE 43a

Brachyscelus globiceps (Claus, 1879)

Female, 8 mm; male, 4 mm

1   Antennula
2   Antenna
3   Mandibula, female
4   Maxilla, female
5   Maxilliped, female
6   Pereiopod 1, male, with detail of propodal spinulation
    (equal in female)

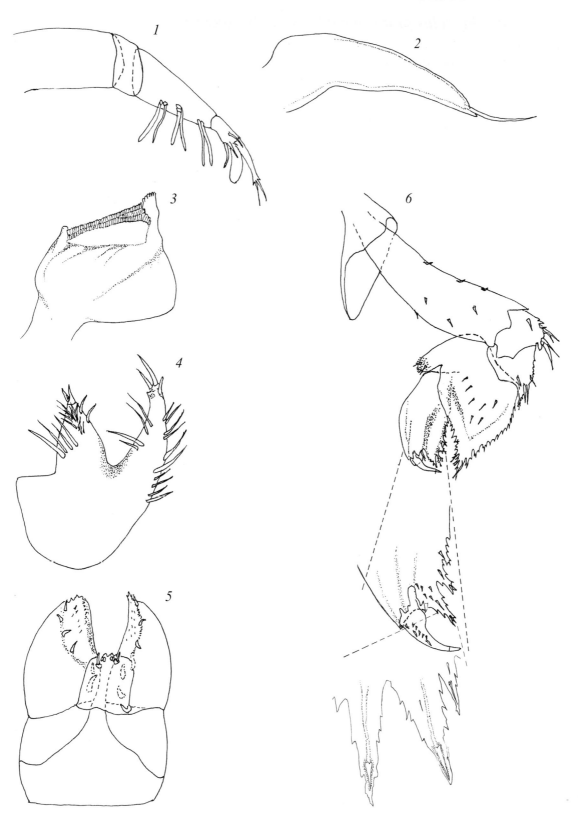

FIGURE 43b

Brachyscelus globiceps (Claus, 1879) (cont.)

7   Pereiopod 2, female
8   Pereiopod 3, female
9   Pereiopod 4, female, with epipodite
10 Pereiopod 5, female

FIGURE 43b

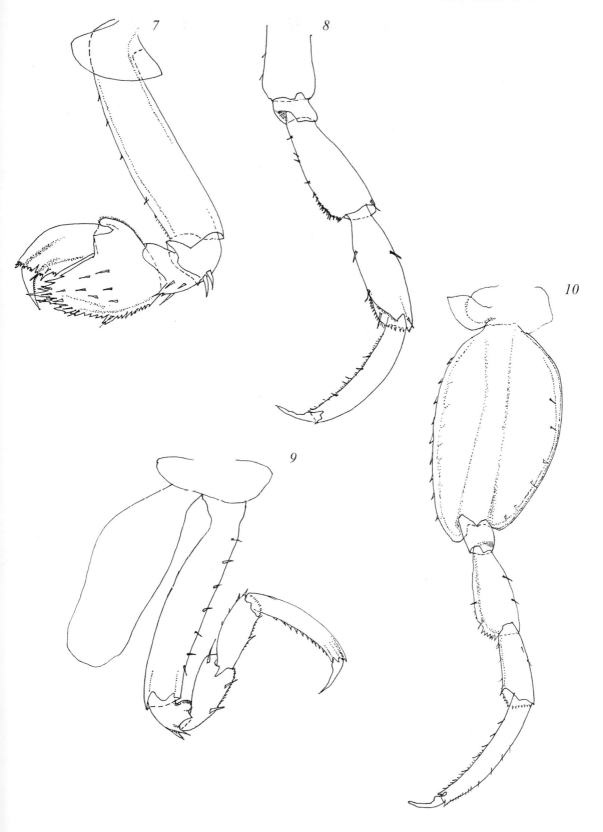

FIGURE 43c

Brachyscelus globiceps (Claus, 1879) (cont.)

11 Pereiopod 6, female, with details
12 Pereiopod 7, female
13 Pleopod 1, female, bifid spine and coupling hooks
14 Uropods and telson, female
15 Uropod 1

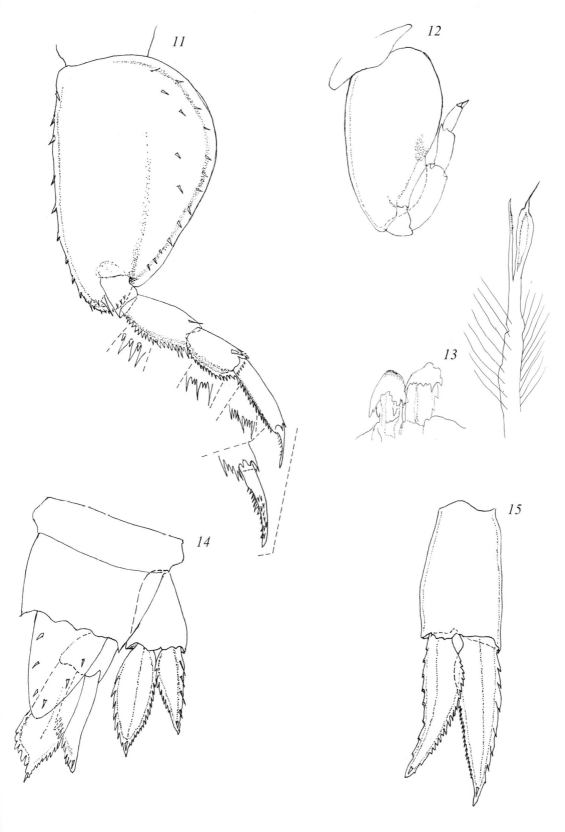

FIGURE 44a

Brachyscelus rapax Claus, 1879

Female, 4 mm

1   Lateral view
2   Antennula
3   Mandibula, gnathobase, partial
4   Maxilliped, dorsal view
5   Maxilliped, lateral view
6   Pereiopod 1, with combined details
7   Pereiopod 2, with detail

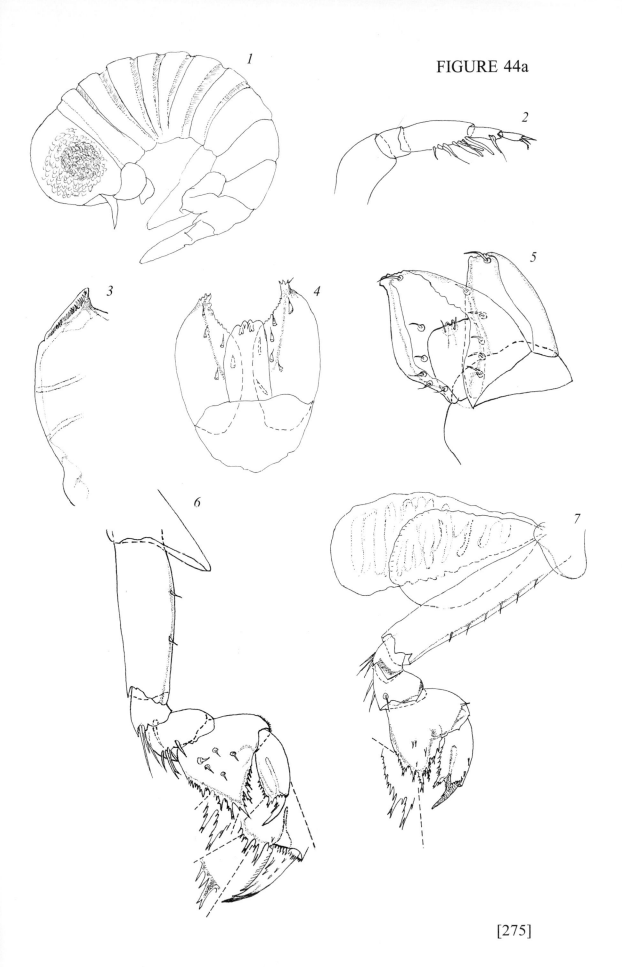

1

2

3

4

5

6

7

FIGURE 44b

Brachyscelus rapax Claus, 1879 (cont.)

8   Pereiopod 3
9   Pereiopod 4
10 Pereiopod 5, with details
11 Pereiopod 6, with details
12 Pereiopod 7, right, with detail
13 Pereiopod 7, left, asymmetrical, with detail
14 Uropods and telson, with detail

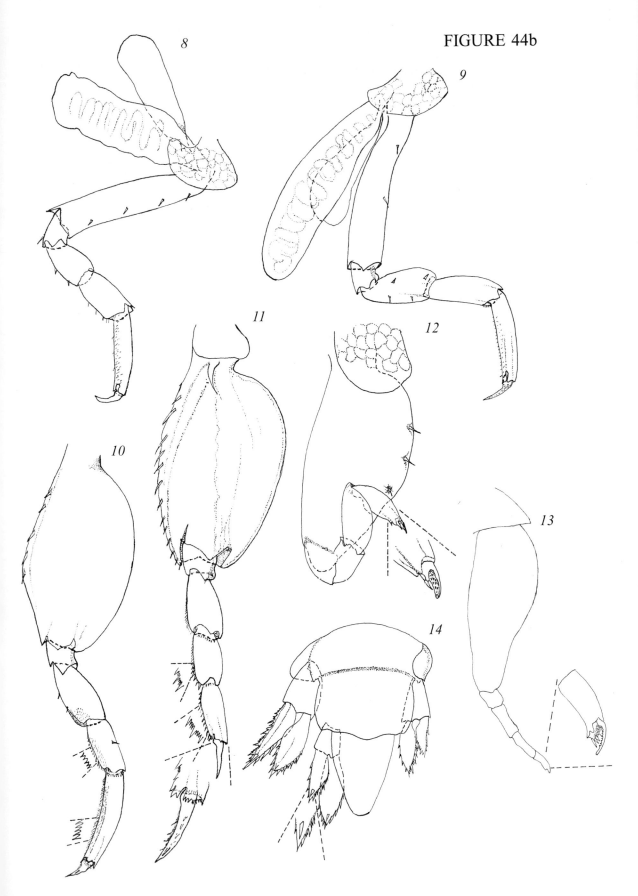

FIGURE 45a

Euthamneus rostratus (Bovallius, 1887)

Females, 3 and 6 mm; male, 3 mm

1   Lateral view
2   Antennula, female
3   Mandibula, gnathobase and palp, male
4   Maxilliped, male
5   Pereiopod 1, female 6 mm, right
6   Pereiopod 1, female 6 mm, left
7   Pereiopod 2, female 6 mm
8   Pereiopod 3, female 3 mm

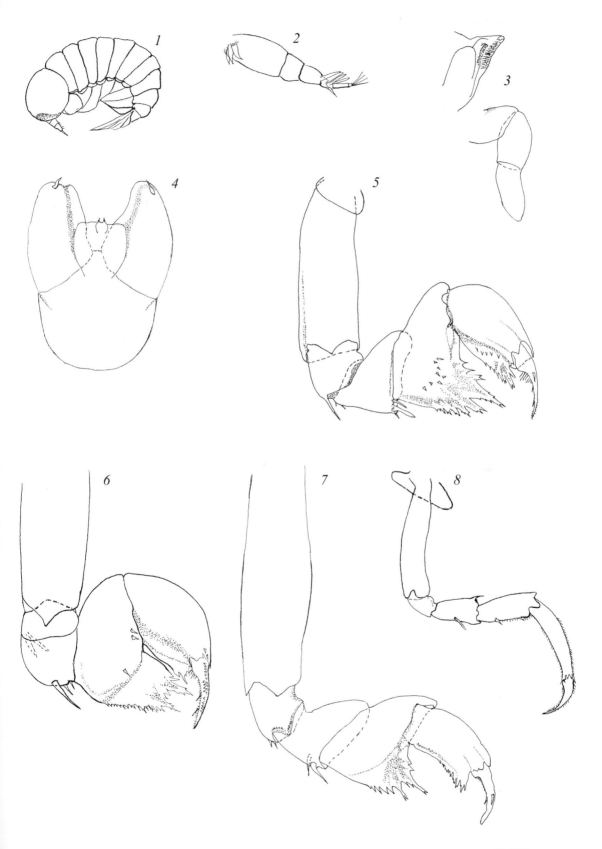

FIGURE 45b

Euthamneus rostratus (Bovallius, 1887) (cont.)

9   Pereiopod 3, male
10 Pereiopod 4, female 3 mm
11 Pereiopod 5, female 6 mm
12 Pereiopod 6, female 6 mm
13 Pereiopod 7, female 6 mm, with detail
14 Uropods and telson, female 6 mm

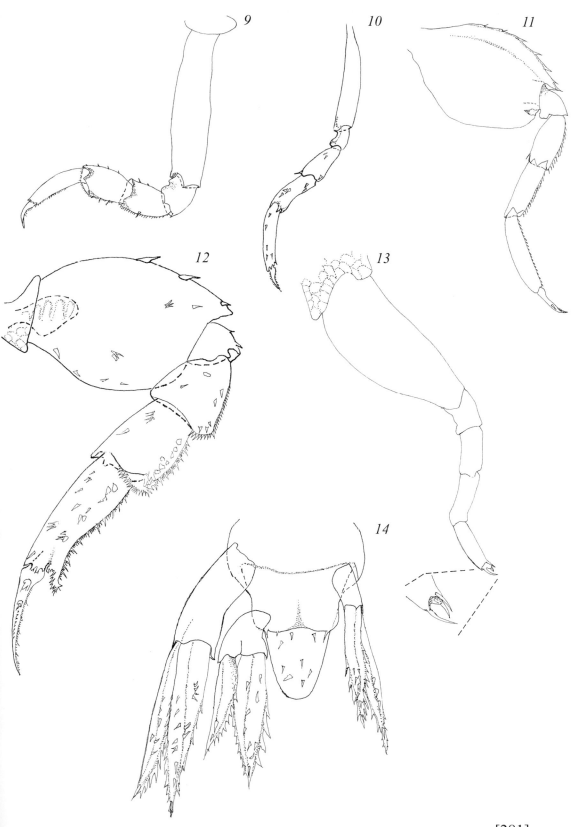

FIGURE 46a

OXYCEPHALIDAE Bate

Oxycephalus piscator Milne Edwards, 1830

Female, 18 mm

1 Lateral view
2 Antennula, with details
3 Detail of edge of antennular cephalic segment
4 Mandibulae, left and right

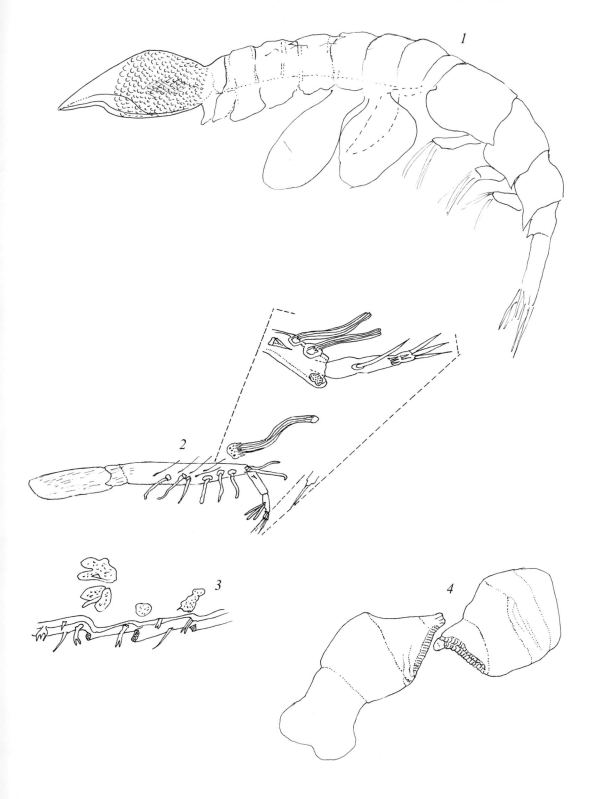

FIGURE 46a

1

2

3

4

FIGURE 46b

Oxycephalus piscator Milne Edwards, 1830 (cont.)

5   Maxilliped, anterior view
6   Pereiopod 1
7   Details of pereiopod 1: dactyl and spines

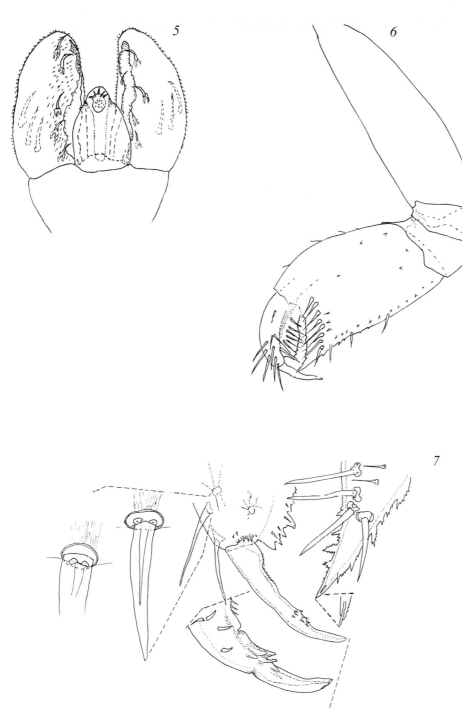

FIGURE 46c

Oxycephalus piscator Milne Edwards, 1830 (cont.)

8 Pereiopod 2, with details
9 Pereiopod 3, with detail of dactyl
10 Pereiopod 4, with detail of dactyl

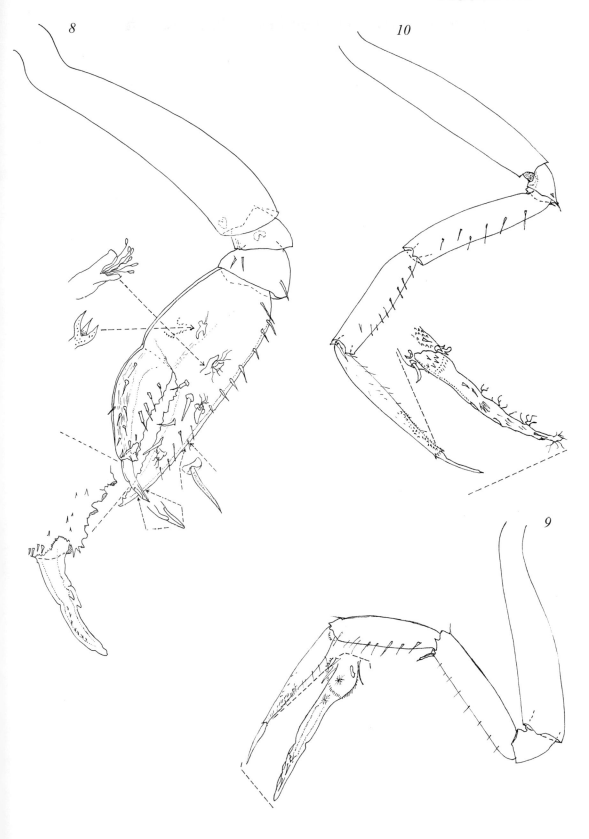

FIGURE 46d

Oxycephalus piscator Milne Edwards, 1830 (cont.)

11 Pereiopod 5, with details
12 Pereiopod 6, with details
13 Pereiopod 7
14 Uropods and telson, with details

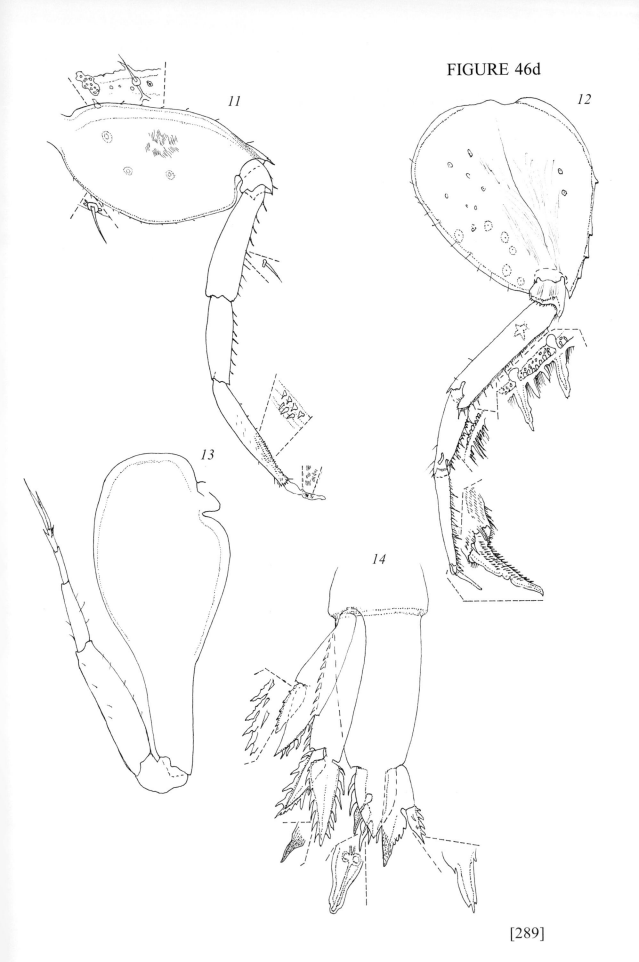

FIGURE 47a

Oxycephalus clausi Bovallius, 1887

Female, 27 mm

1 Head and thorax outline, lateral view
2 Pereiopod 1
3 Pereiopod 2
4 Pereiopod 5

FIGURE 47a

FIGURE 47b

Oxycephalus clausi Bovallius, 1887 (cont.)

5 Pereiopod 6
6 Pereiopod 7
7 Uropod 1, with detail
8 Uropod 2, with detail
9 Uropod 3 and telson
10 Last epimera

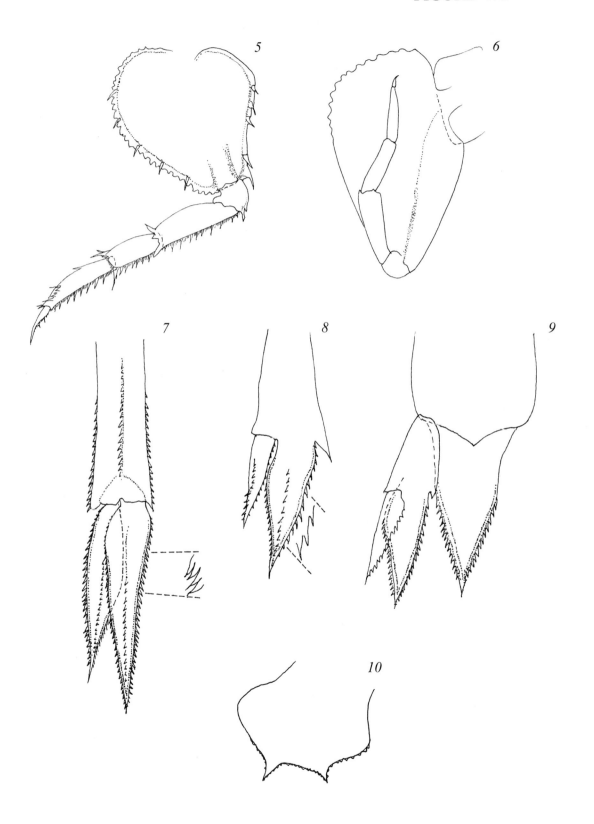

FIGURE 48a

Streetsia challengeri Stebbing, 1888

Female, 19 mm; males, 8, 14 and 25 mm

1   Lateral view
2   Antennula, male 14 mm, with details
3   Antennula, male 25 mm, with detail of tubuliform spine
4   Antenna, female

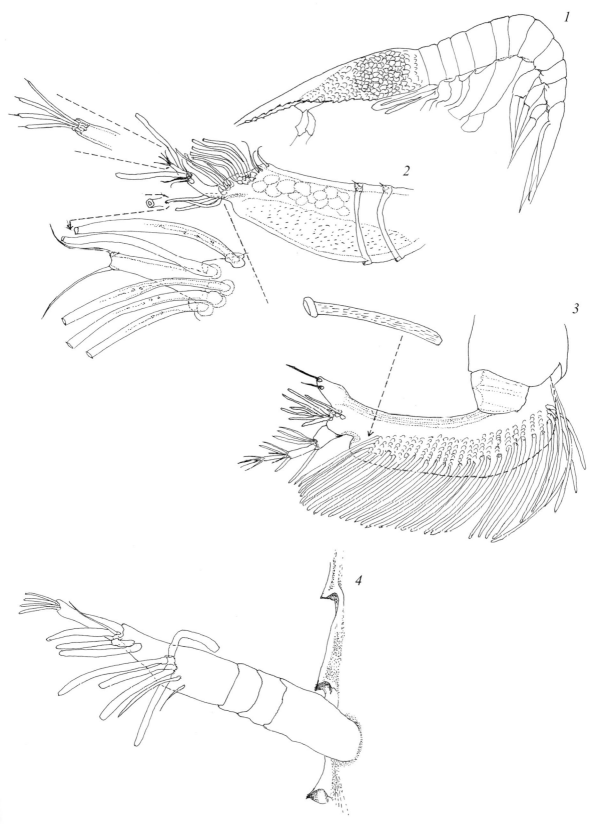

FIGURE 48b

Streetsia challengeri Stebbing, 1888 (cont.)

5 Antenna, male
6 Mandibula, female
7 Mandibula, male 25 mm, with palp
8 Rudimentary maxilla, female
9 Maxilliped, female
10 Maxilliped, male, upper border of external lobe
11 Pereiopod 1, female, with details

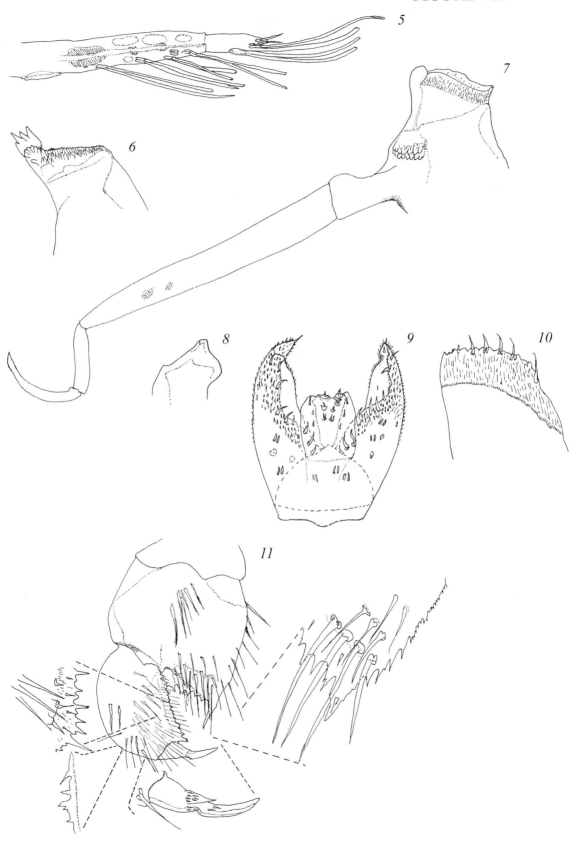

FIGURE 48c

Streetsia challengeri Stebbing, 1888 (cont.)

12 Pereiopod 1, female, large spines on surface
13 Pereiopod 1, male 14 mm, with details
14 Pereiopod 1, male 25 mm, dactyl and fragment
    from segment 5
15 Pereiopod 2, female, with detail

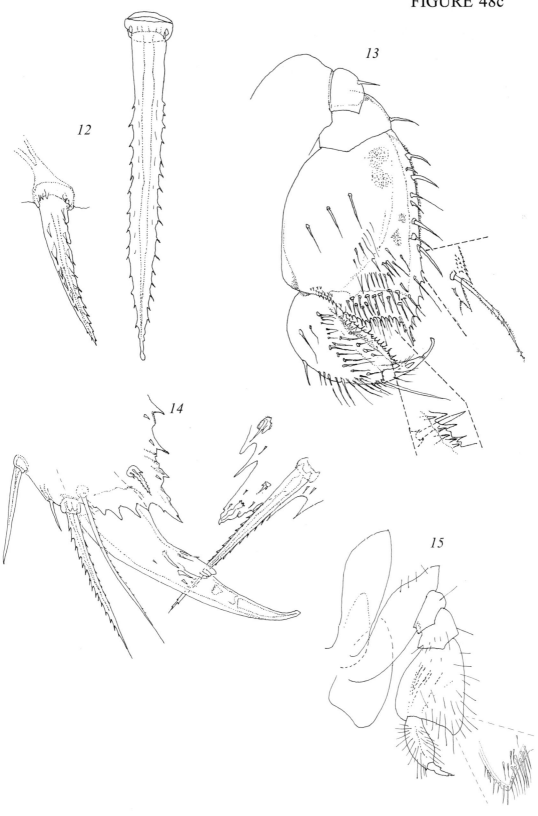

FIGURE 48d

Streetsia challengeri Stebbing, 1888 (cont.)

16 Pereiopod 2, female, dactyl and large propodial spine
17 Pereiopod 2, male, with details
18 Pereiopod 3, female
19 Pereiopod 4, female, with detail
20 Pereiopod 5, female, with details

FIGURE 48e

Streetsia challengeri Stebbing, 1888 (cont.)

21 Pereiopod 6, female
22 Pereiopod 6, female, last three segments with details
23 Pereiopod 7, female
24 Pleopod 1, basal part of exopodite
25 Pleopod 1, male, coupling hooks
26 Last pleonal epimere
27 Uropods and telson, with details

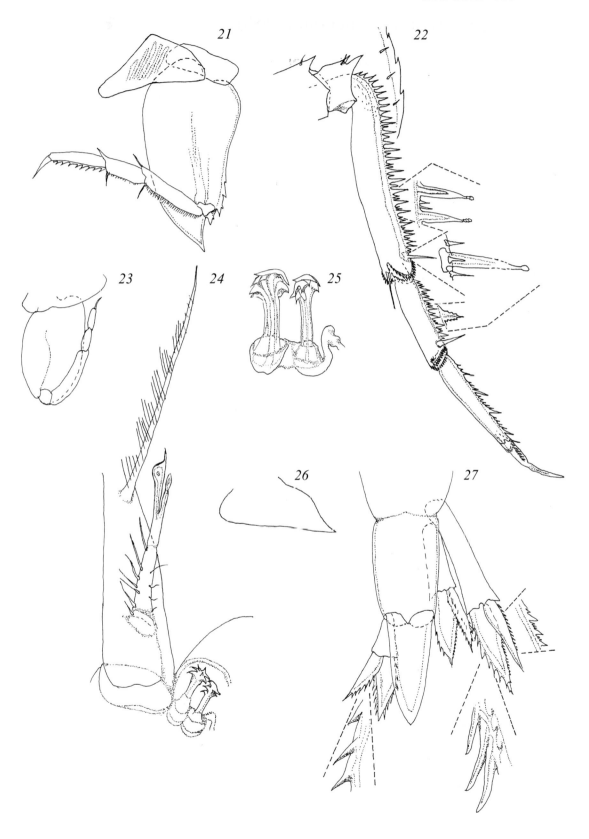

FIGURE 49a

Streetsia steenstrupi (Bovallius, 1887)

Female, 4 mm; male (probably immature), 7 mm

1 Male, lateral view
2 Head of female, view from above
3 Pereiopod 1, male
4 Pereiopod 2, male
5 Pereiopod 3, female, with a possible receptor
6 Pereiopod 4, male

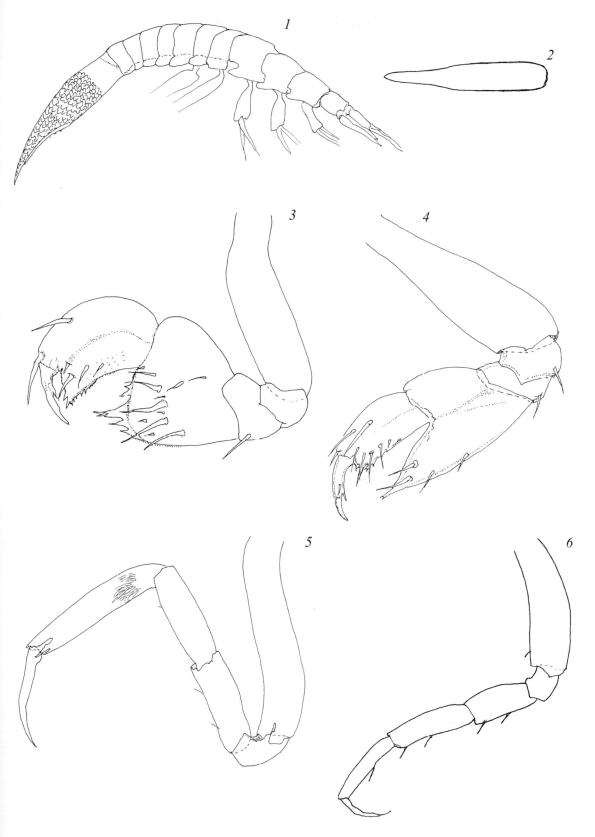

FIGURE 49b

Streetsia steenstrupi (Bovallius, 1887) (cont.)

7 Pereiopod 5, male
8 Pereiopod 6, female, with details
9 Pereiopod 6, male
10 Pereiopod 7, male
11 Coupling hooks of pleopod 1
12 Uropods and telson, with details

FIGURE 49b

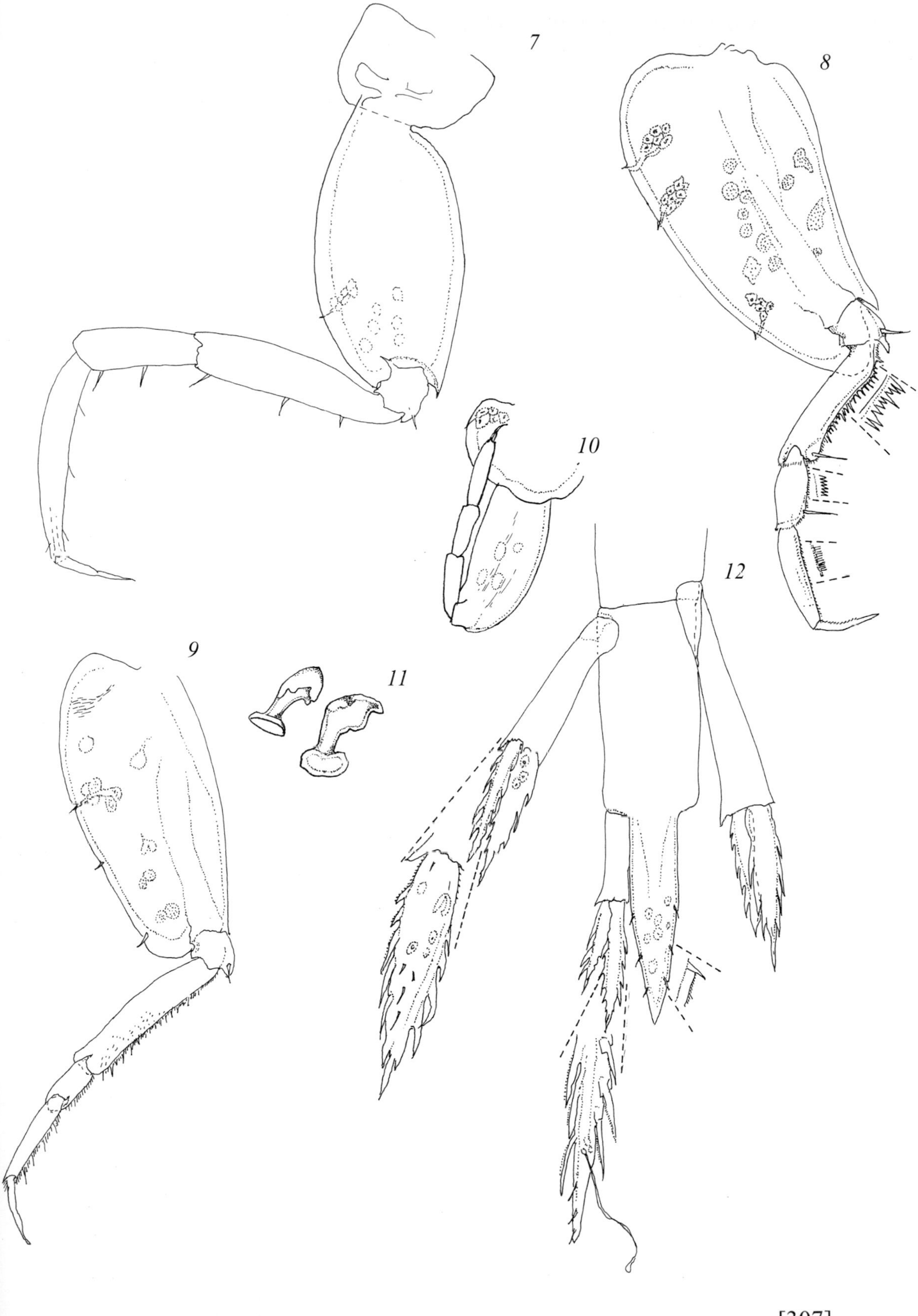

FIGURE 50a

Streetsia porcella (Claus, 1879)

Female (?), 12 mm

1 Lateral view
2 Pereiopod 1
3 Pereiopod 2, with detail
4 Pereiopod 3
5 Pereiopod 4

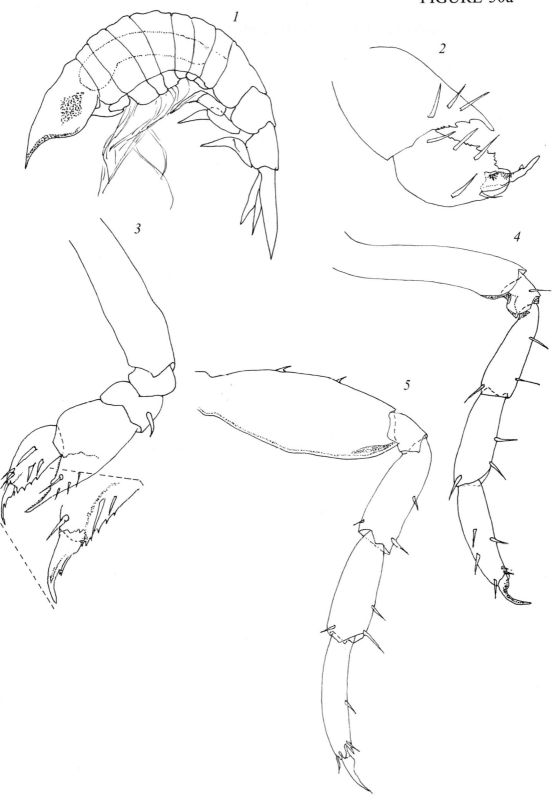

FIGURE 50a

FIGURE 50b

Streetsia porcella (Claus, 1879) (cont.)

6 Pereiopod 5
7 Pereiopod 6, with details
8 Pereiopod 7, with detail
9 Pleonal epimeres 2 and 3
10 Uropods and telson, with details

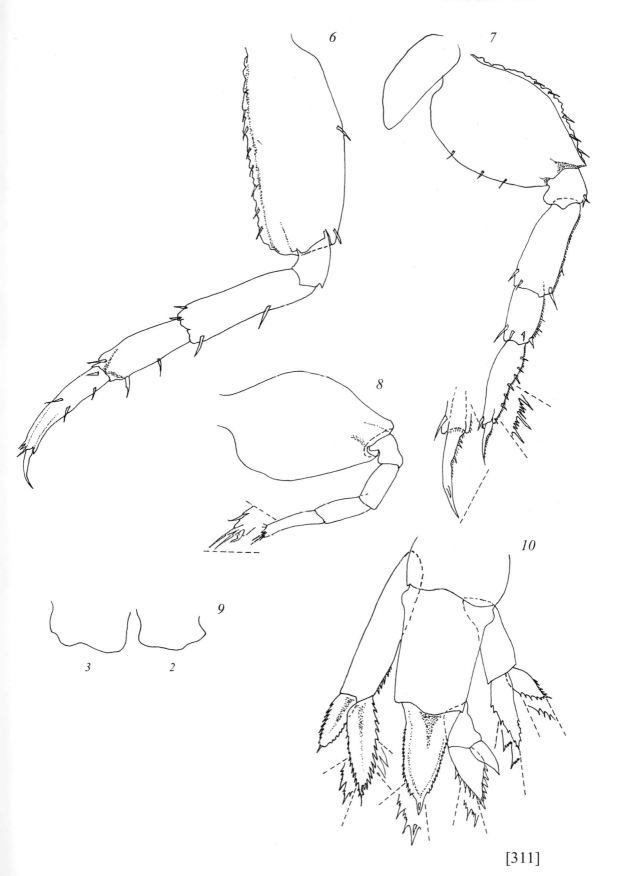

FIGURE 51a

Leptocotis tenuirostris (Claus, 1871)

Female, 17 mm; male, 18 mm

1   Female, lateral view
2   Male, head in lateral view
3   Antennula, female
4   Antennula, male, with details
5   Maxilliped, female
6   Maxilla, female
7   Mandibula, female

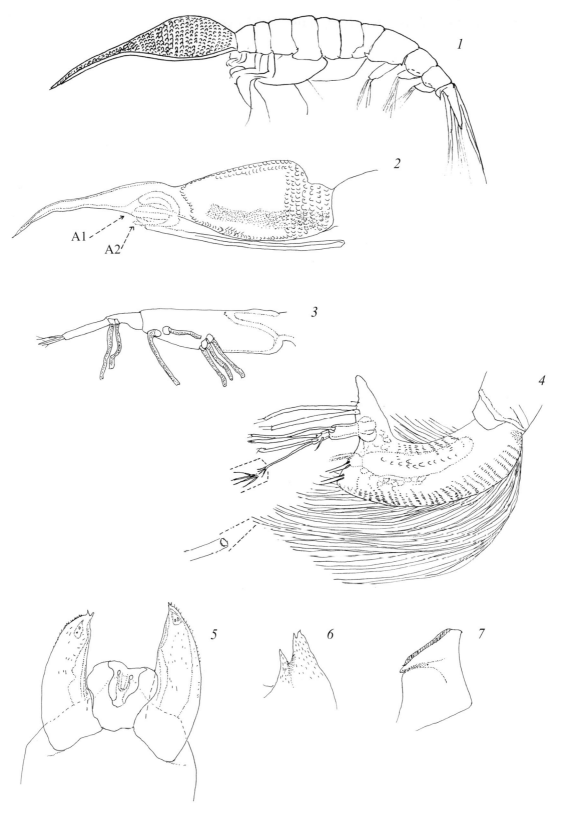

FIGURE 51b

Leptocotis tenuirostris (Claus, 1871) (cont.)

8 Pereiopod 1, female
9 Pereiopod 1, male
10 Pereiopod 2, male
11 Pereiopod 3, male
12 Pereiopod 4, male
13 Pereiopod 5, male

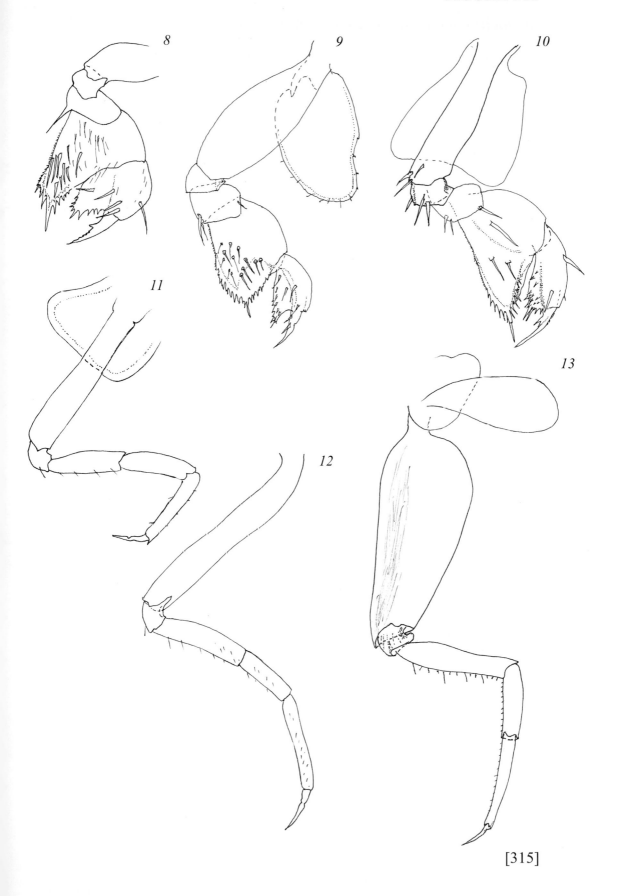

FIGURE 51c

Leptocotis tenuirostris (Claus, 1871) (cont.)

14 Pereiopod 6, male, with details
15 Pereiopod 7, male
16 Pleonal epimeres 1 and 2
17 Uropods and telson, male, with details
18 Uropod 2, male

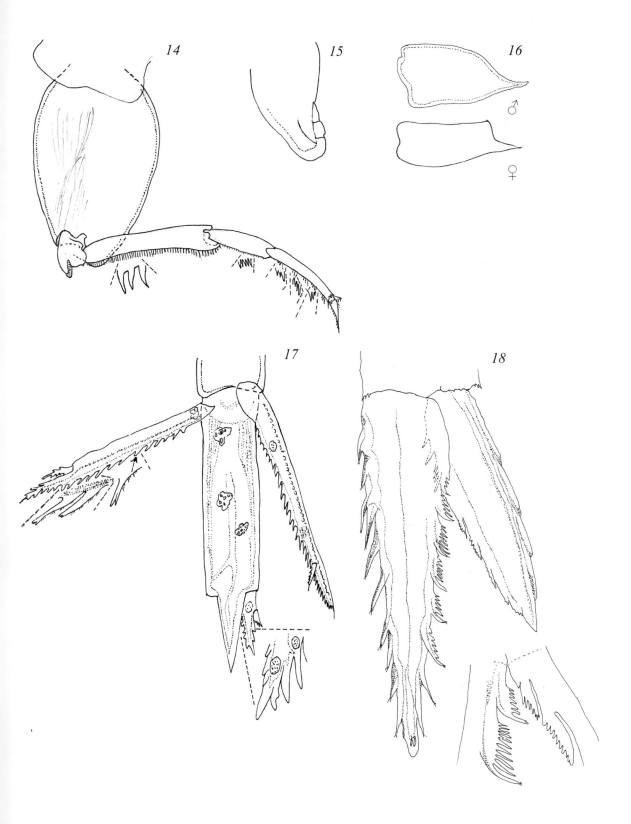

FIGURE 52a

Calamorhynchus pellucidus Streets, 1878

Female, 12 mm; male, 13 mm

1   Female, lateral view
2   Head, from above
3   Head, from below
4   Head, lateral view
5   Antenna, male, with details
6   Molar plate of mandibula, male
7   Maxilliped, female

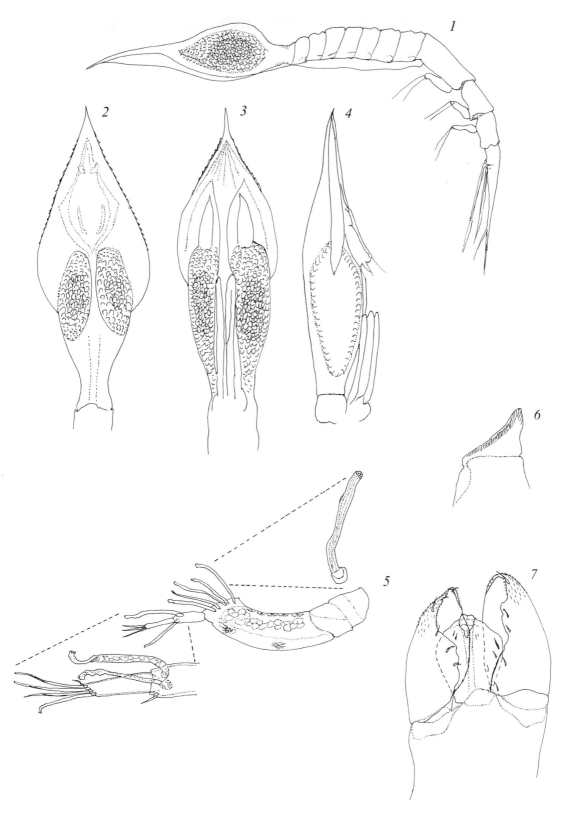

FIGURE 52b

Calamorhynchus pellucidus Streets, 1878 (cont.)

8   Pereiopod 1, female
9   Pereiopod 1, male
10 Pereiopod 2, female, with details
11 Pereiopod 2, male
12 Pereiopod 3, female, with details

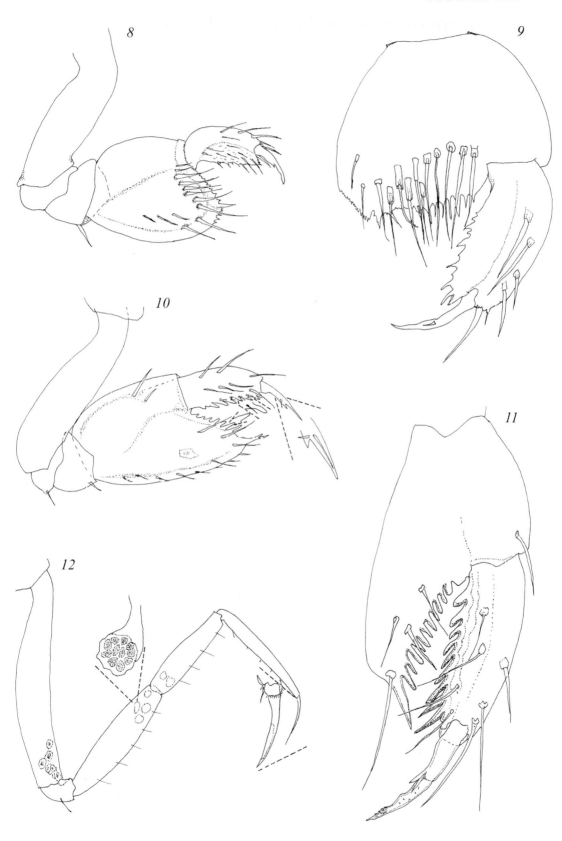

FIGURE 52c

Calamorhynchus pellucidus Streets, 1878 (cont.)

13 Pereiopod 4, female
14 Pereiopod 5, female
15 Pereiopod 6, female, with detail
16 Pereiopod 7, female
17 Last epimere
18 Uropods and telson, with details

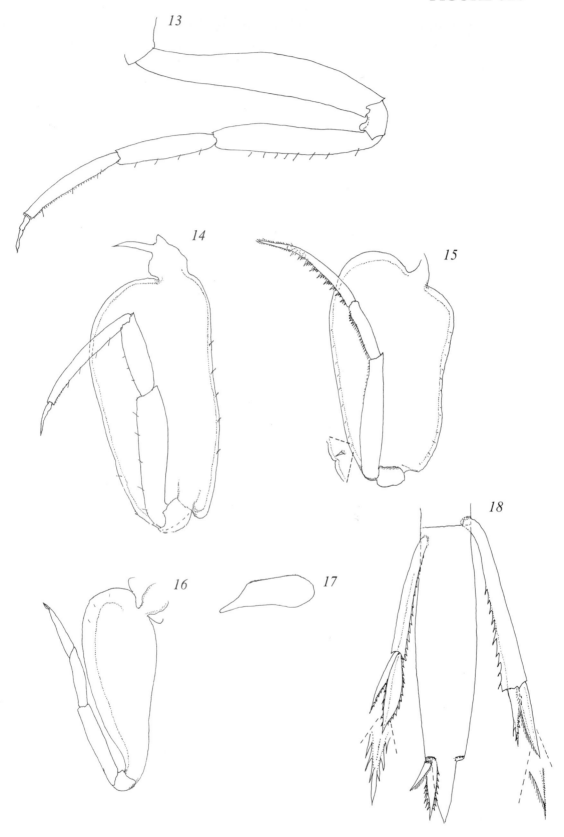

FIGURE 53a

Glossocephalus milne edwardsi Bovallius, 1887

Female, 8 mm

1 Lateral view, schematic
2 Maxilla
   (note: mandibula and maxillula probably completely reduced)
3 Maxilliped, fragment; outer and inner lobes
4 Pereiopod 1
5 Pereiopod 2

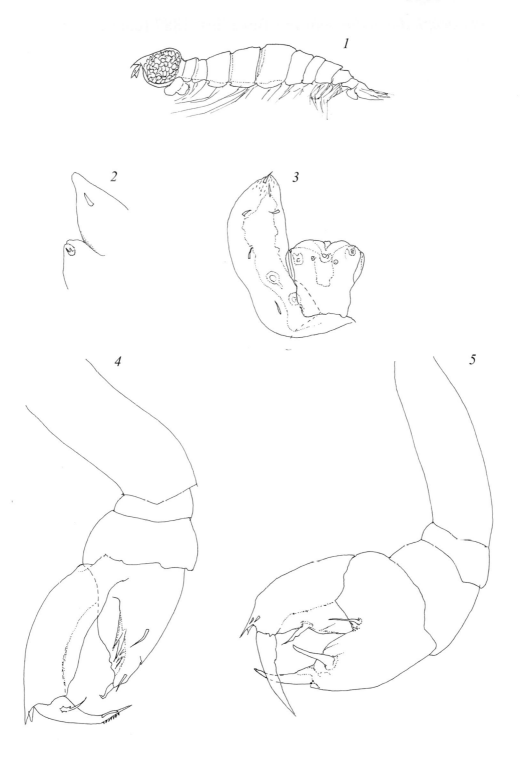

FIGURE 53b

Glossocephalus milne edwarsi Bovallius, 1887 (cont.)

6 Pereiopod 3
7 Pereiopod 5, with detail
8 Pereiopod 6, with detail
9 Pereiopod 7
10 Uropods and telson, with detail

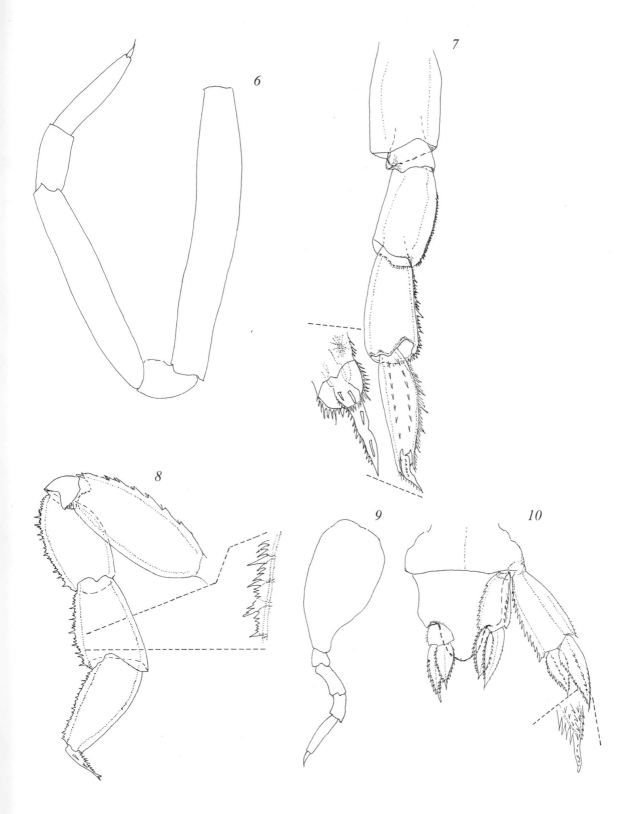

FIGURE 54a

Cranocephalus scleroticus (Streets, 1878)

Female, 8 mm

1   Lateral view
2   Mandibulae, left and right
3   Maxilliped, with detail
4   Pereiopod 1, with detail

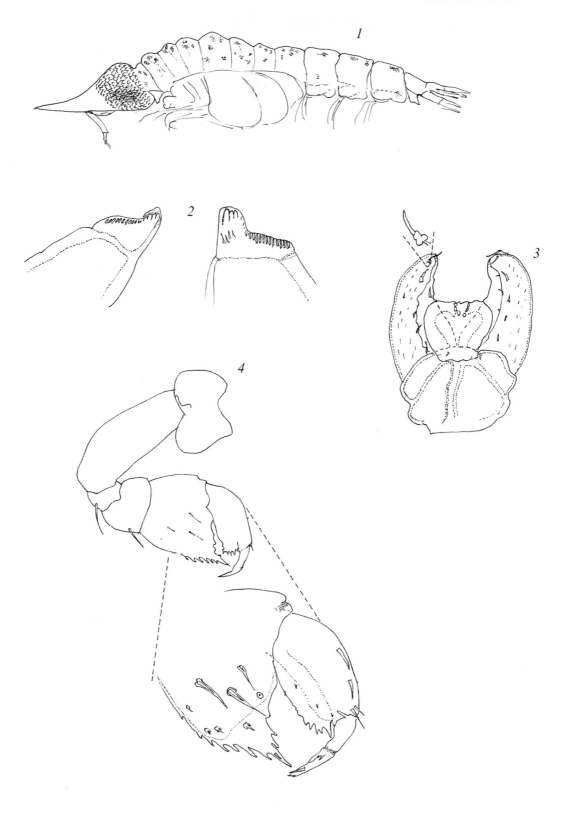

FIGURE 54b

Cranocephalus scleroticus (Streets, 1878)(cont.)

5 Pereiopod 2, with detail
6 Pereiopod 3, with detail
7 Pereiopod 4
8 Pereiopod 5, with details of pores and "cells"

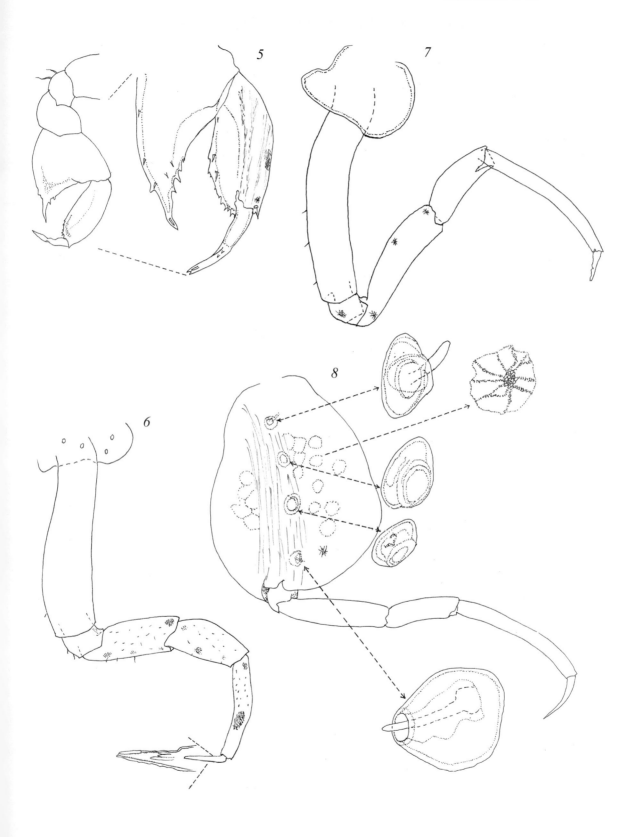

FIGURE 54c

Cranocephalus scleroticus (Streets, 1878) (cont.)

9   Pereiopod 6, with details of pores and "cells"
10 Pereiopod 7, with details of pores
11 Uropods and telson, with details

FIGURE 54c

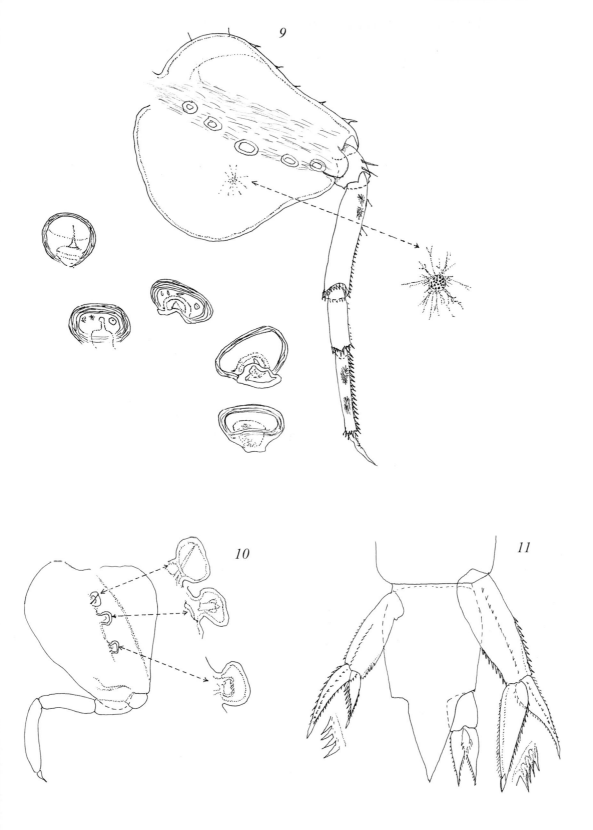

FIGURE 55a

Rhabdosoma whitei Bate, 1862

Male 70 mm and male exuvium 68 mm

1 Lateral view
2 Basal part of head, ventral view
3 Ventral keel of head with bases of antennula and antenna
4 Mandibular complex
5 Inferolateralia in foregut
6 Maxilliped

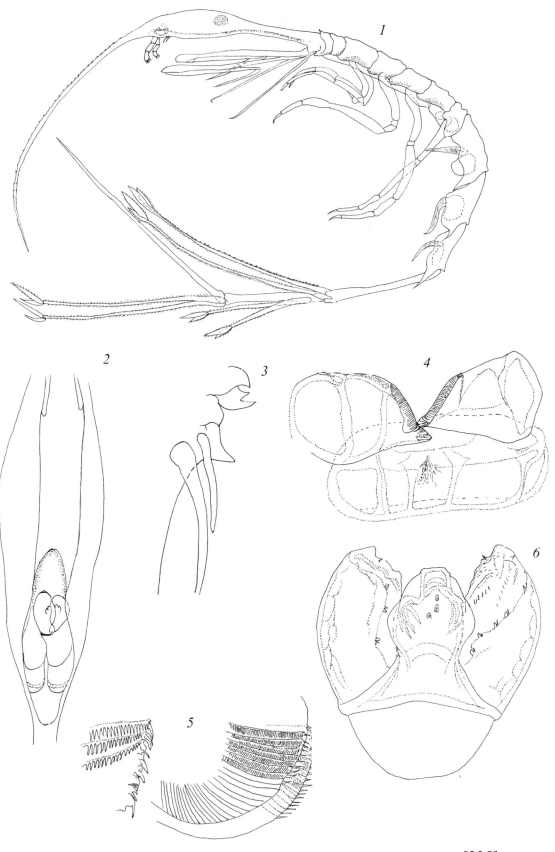

FIGURE 55b

Rhabdosoma whitei Bate, 1862 (cont.)

7 Pereiopod 1
8 Pereiopod 2
9 Pereiopod 5
10 Pereiopod 6, with details
11 Uropod 1
12 Uropod 3
13 Telson
14 Penultimate epimere and epipod

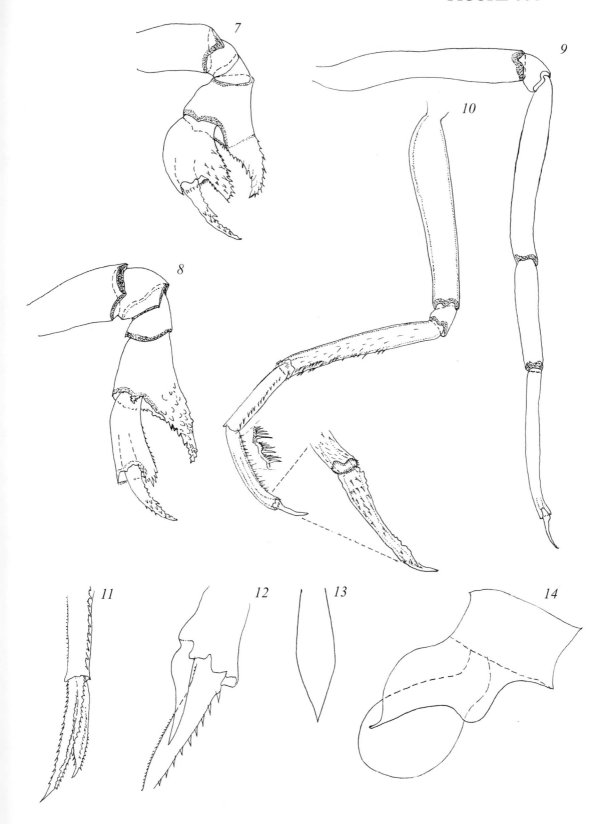

FIGURE 56a

PLATYSCELIDAE Bate

Platyscelus ovoides (Risso, 1816)

Females 9, 13 and 18 mm

1 Lateral view, female 9 mm
2 Lateral view, female 18 mm
3 Mandibula, female 13 mm
4 Mandibulae left and right, female 18 mm
5 Maxillula, with details
6 Maxilla, tips of palp

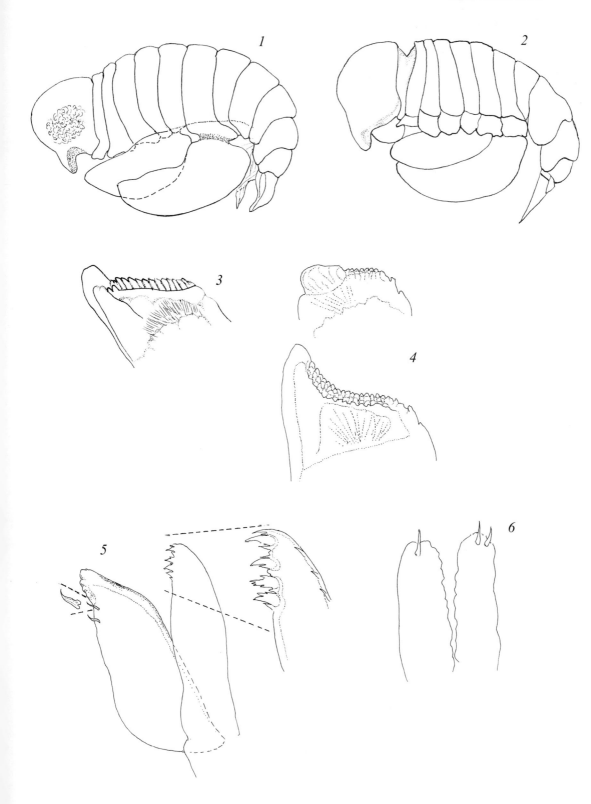

FIGURE 56b

Platyscelus ovoides (Risso, 1816) (cont.)

7   Maxilliped, female 13 mm, with details
8   Maxilliped, female 18 mm, with details
9   Pereiopod 1, with details
10 Pereiopod 2

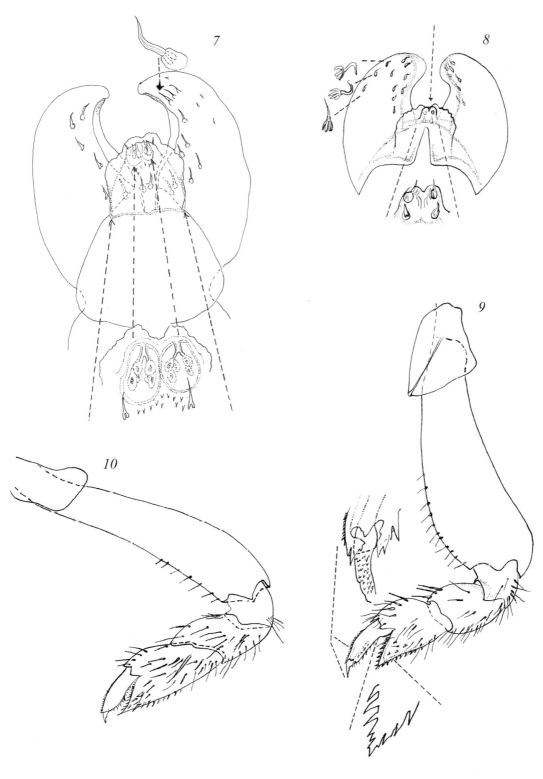

FIGURE 56c

Platyscelus ovoides (Risso, 1816) (cont.)

11 Pereiopod 3
12 Pereiopod 4, with detail
13 Pereiopod 5, with details
14 Pereiopod 6, with details

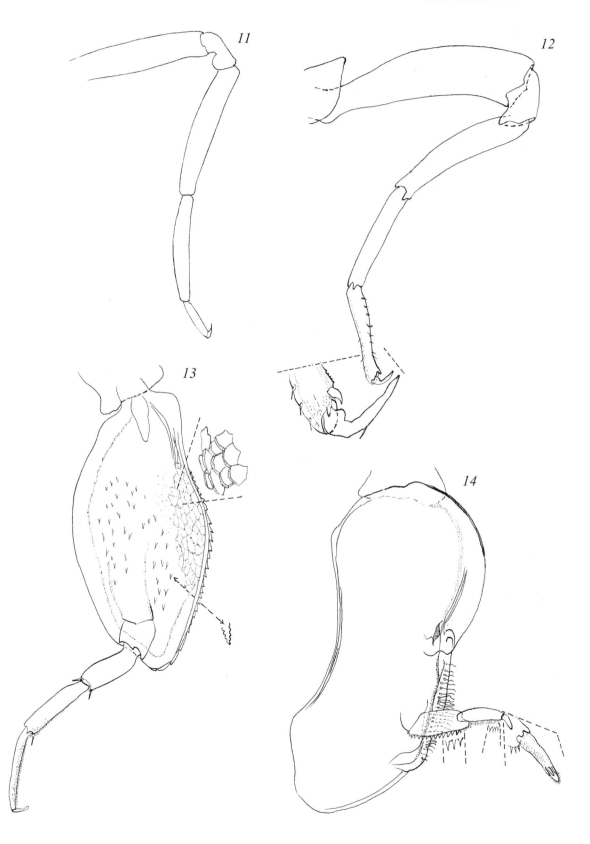

FIGURE 56d

Platyscelus ovoides (Risso, 1816) (cont.)

15 Pereiopod 7
16 Bifid spines on pleopods 1, 2 and 3, female 9 mm
17 Bifid spine on pleopod 1, female 13 mm
18 Coupling hooks of pleopods 1, 2, female 9 mm
19 Coupling hooks of pleopods 1, 2, female 13 mm
20 Coupling hooks of pleopods 1, 2, female 18 mm
21 Uropods and telson, juvenile female 9 mm
22 Uropods and telson, female 18 mm

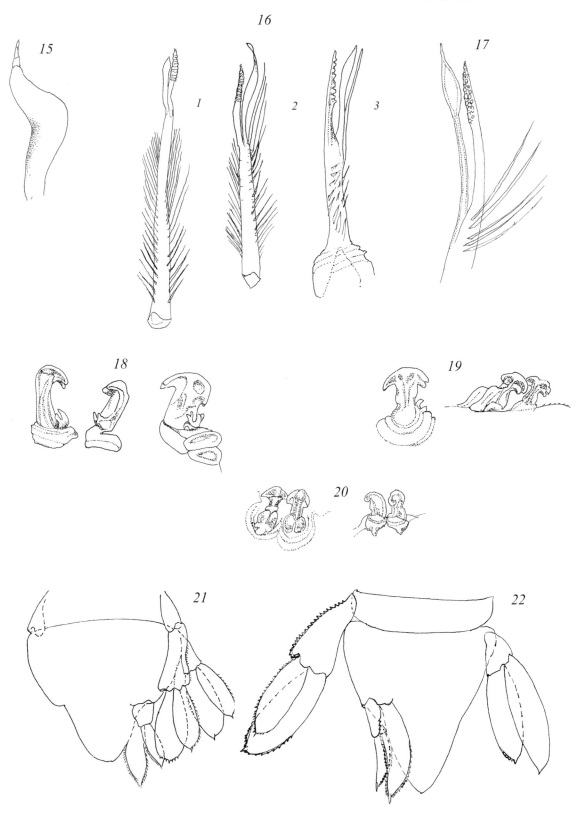

FIGURE 57a

Platyscelus armatus (Claus, 1879)

1   Antennula
2   Mandibula
3   Maxillula, palp, outer lobe reduced
4   Maxilliped, posterior view, without right lobe
5   Pereiopod 1, with details
6   Pereiopod 2, with detail

FIGURE 57b

Platyscelus armatus (Claus, 1879) (cont.)

7   Pereiopod 3, with detail
8   Pereiopod 4
9   Pereiopod 5

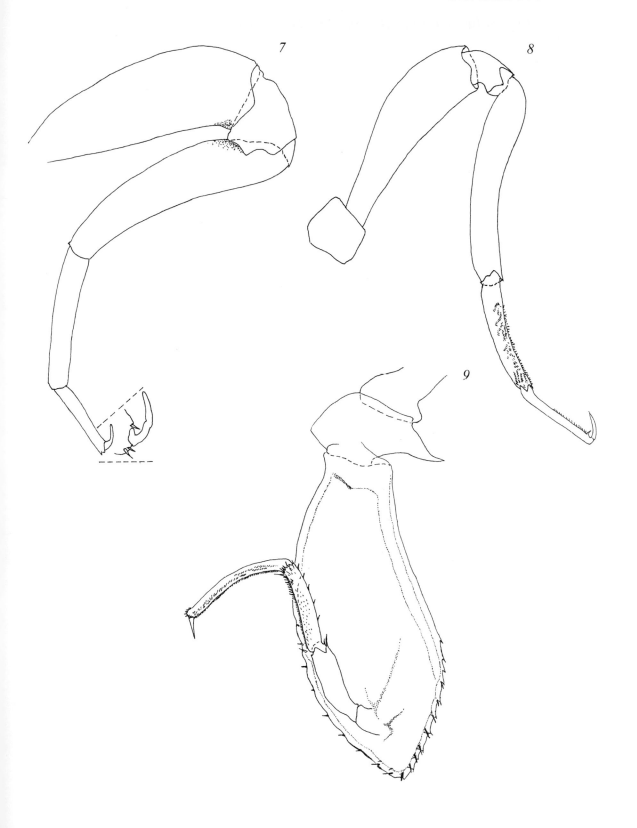

FIGURE 57c

Platyscelus armatus (Claus, 1879) (cont.)

10 Pereiopod 6, with details
11 Pereiopod 7
12 Uropod 1
13 Uropod 2
14 Uropod 3

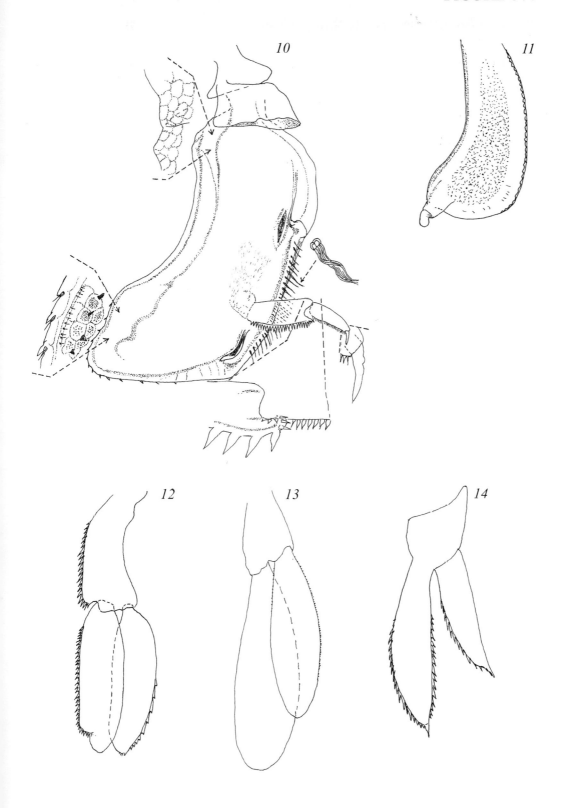

FIGURE 58a

Platyscelus serratulus Stebbing, 1888

Females, 5 and 7 mm; males, 5 and 7 mm

1   Lateral view, female
2   Antennula, male
3   Antenna, male
4   Mandibulae, male, left and right, with detail

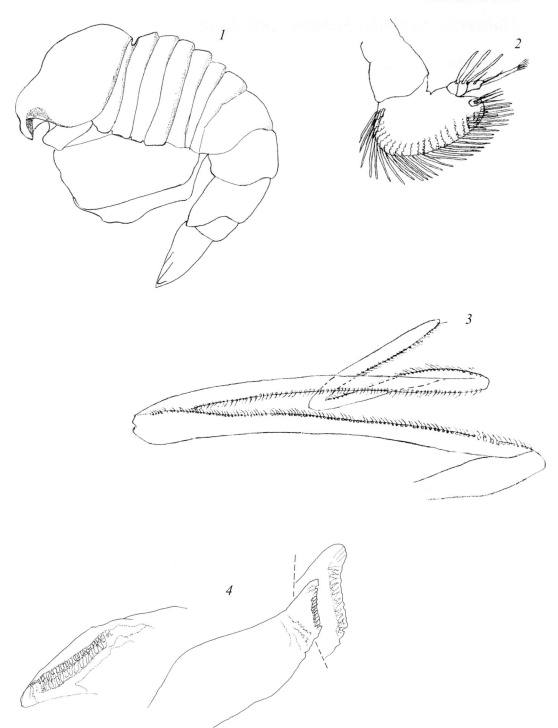

FIGURE 58a

FIGURE 58b

Platyscelus serratulus Stebbing, 1888 (cont.)

5 Mandibula, male, with palp
6 Maxilliped, female
7 Maxilliped, male 7 mm, with details of inner lobe
8 Maxilliped, male

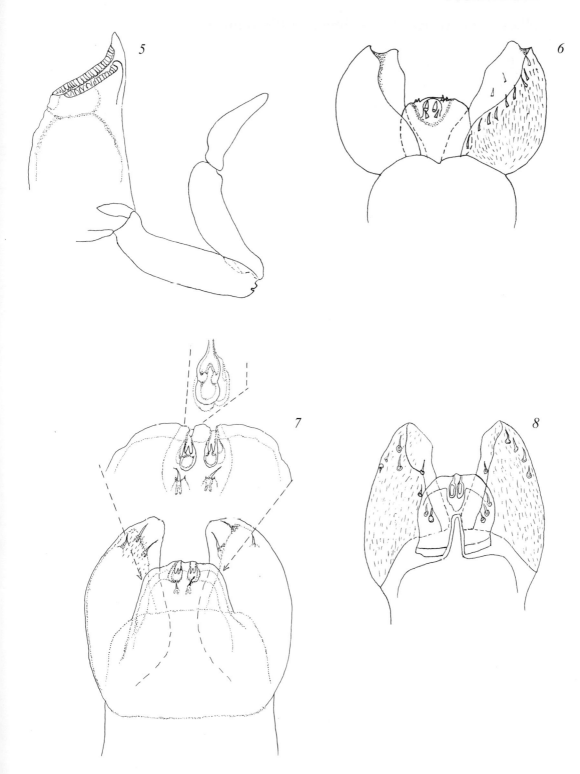

FIGURE 58c

Platyscelus serratulus Stebbing, 1888 (cont.)

9   Pereiopod 1, female
10 Pereiopod 1, male 7 mm, with detail
11 Pereiopod 2, female, with detail
12 Pereiopod 2, male 7 mm, with details
13 Pereiopod 5, male, with details of cuticle

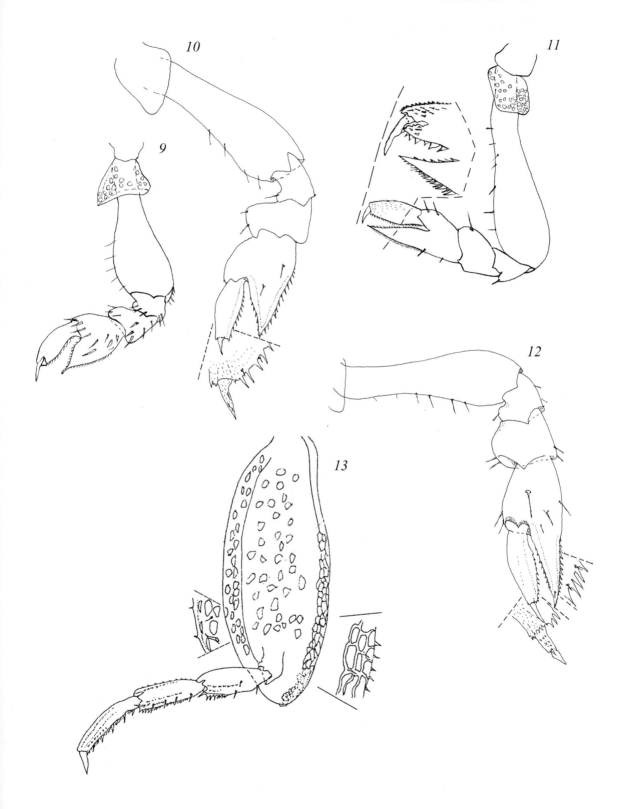

FIGURE 58d

Platyscelus serratulus Stebbing, 1888 (cont.)

14 Pereiopod 6, female, with details
15 Pereiopod 6, male 7 mm
16 Pereiopod 7, female
17 Bifid spine and coupling hooks of pleopod 1
18 Uropods 1, 2, and 3

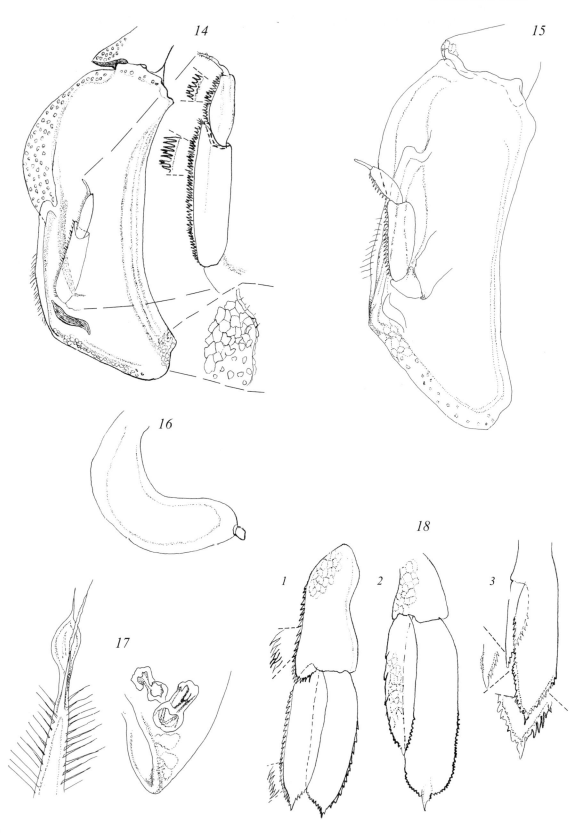

FIGURE 59a

Paratyphis maculatus Claus, 1879

Male, 4.5 mm

1   Antennula
2   Antenna
3   Mandibulae, with palp
4   Maxillula, palp
5   Maxilliped, lateral view, outer lobe

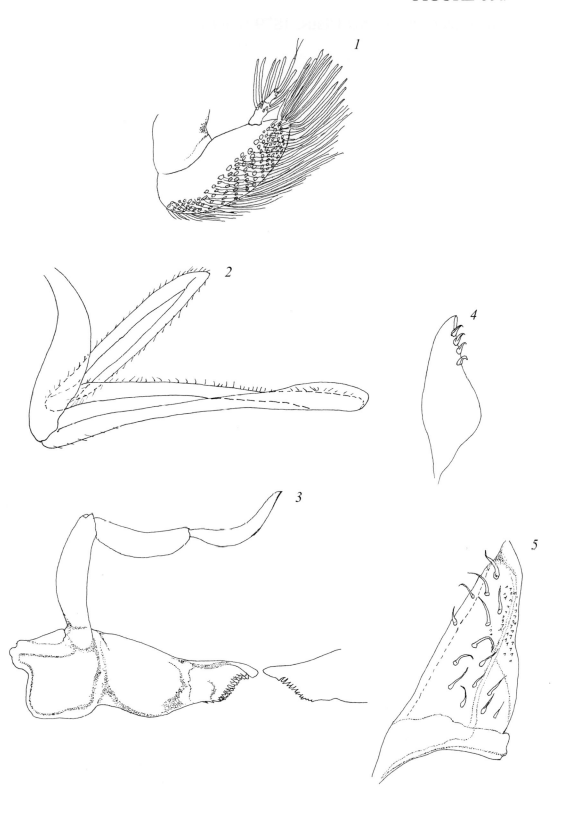

FIGURE 59b

Paratyphis maculatus Claus, 1879 (cont.)

6   Pereiopod 1, with detail
7   Pereiopod 2, with detail
8   Pereiopod 3, with detail

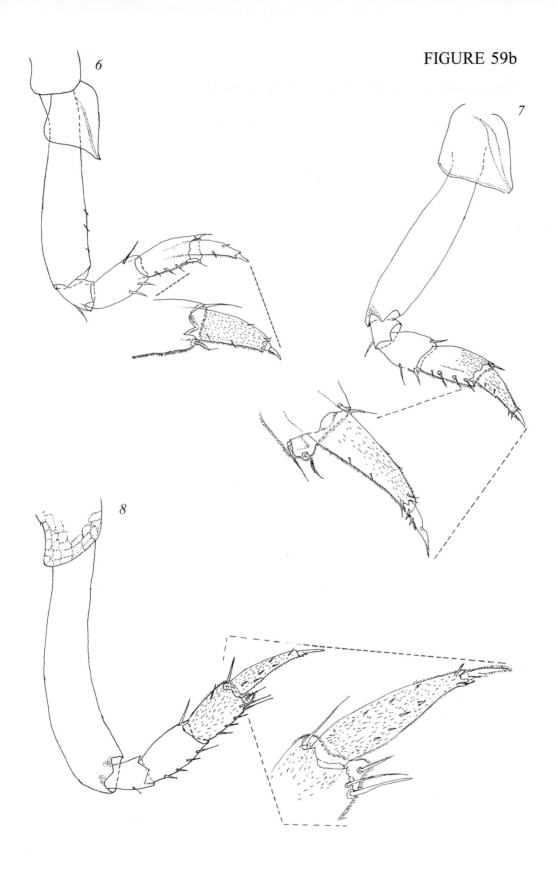

FIGURE 59c

Paratyphis maculatus Claus, 1879 (cont.)

9 Pereiopod 5
10 Pereiopod 6, with detail
11 Pereiopod 7
12 Uropods and telson

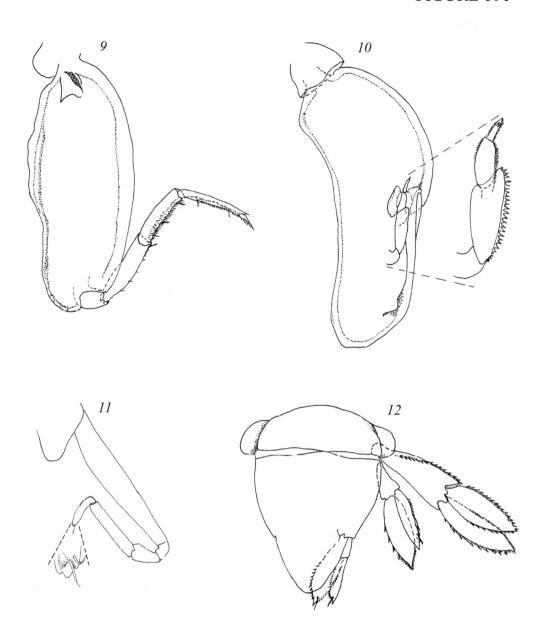

FIGURE 60a

Paratyphis spinosus Spandl, 1924

Female, 4 mm

1   Lateral view
2   Antennula
3   Antenna
4   Mandibula
5   Palp of reduced maxillula
6   Maxilliped
7   Pereiopod 1
8   Pereiopod 2

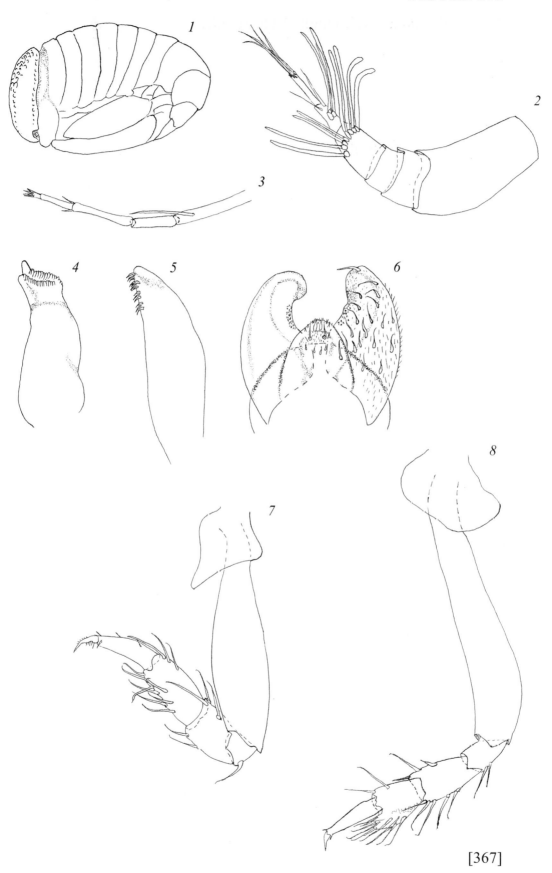

FIGURE 60b

Paratyphis spinosus Spandl, 1924 (cont.)

9   Pereiopod 3
10 Pereiopod 5
11 Pereiopod 6, with detail
12 Pereiopod 7
13 Uropods and telson

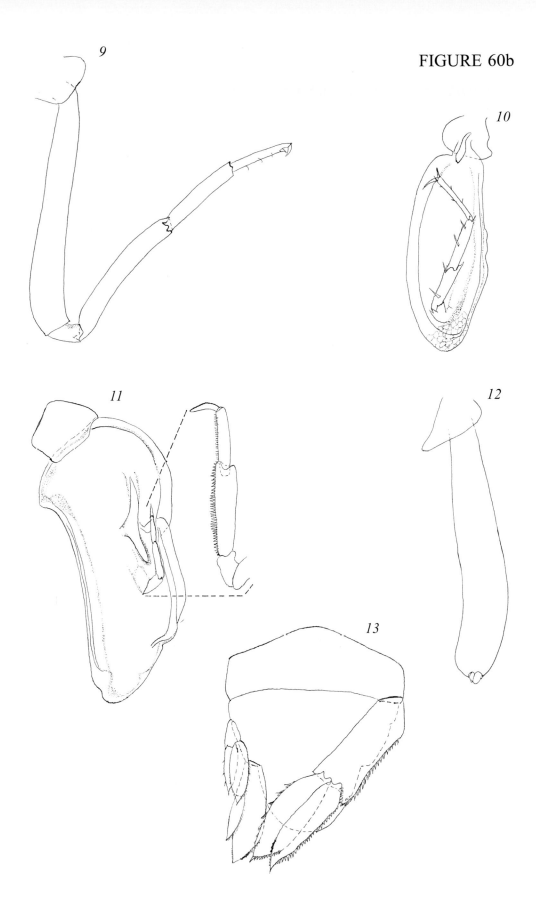

FIGURE 61a

Paratyphis promontorii Stebbing, 1888

Females, 2.5–3 mm; males, 3 and 5 mm

1   Lateral view
2   Antennula, male
3   Antenna, male
4   Mandibula and palp, male
5   Maxillula, male
6   Maxilla, male
7   Maxilliped, male

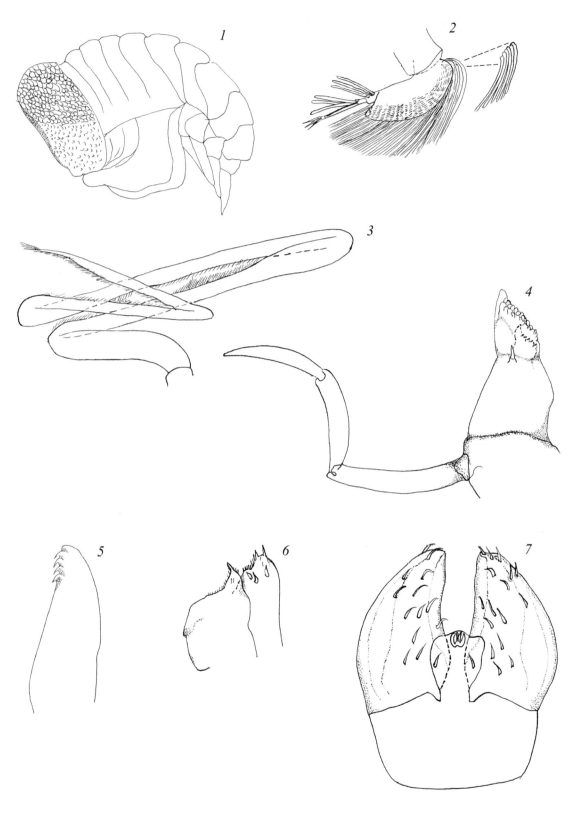

FIGURE 61b

Paratyphis promontorii Stebbing, 1888 (cont.)

8   Pereiopod 1, female
9   Pereiopod 1, male
10 Pereiopod 2, female
11 Pereiopod 2, male, with detail

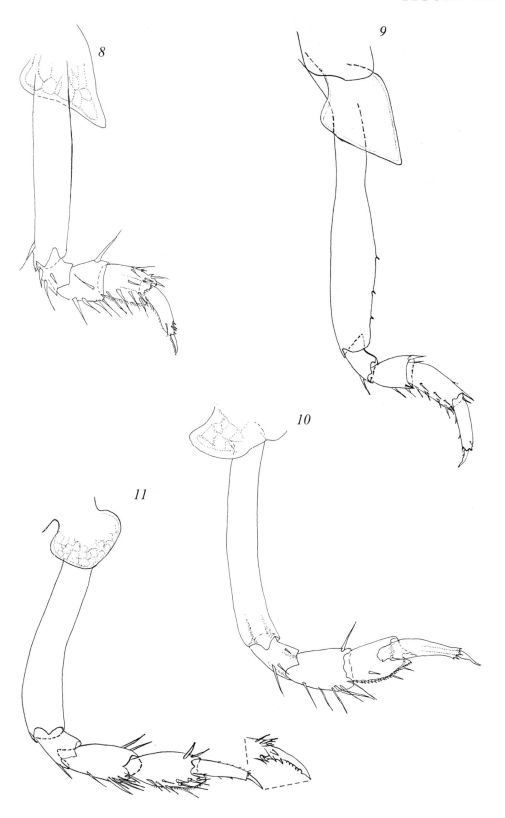

FIGURE 61c

Paratyphis promontorii Stebbing, 1888 (cont.)

12 Pereiopod 3, female, with detail
13 Pereiopod 3, male, with detail
14 Pereiopod 4, female
15 Pereiopod 4, male, with detail of right and left
    asymmetrical dactyli

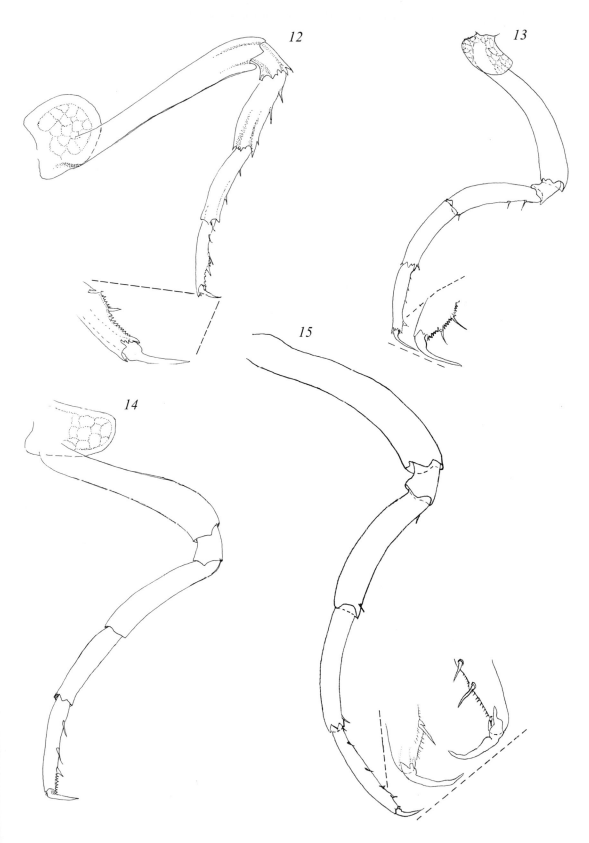

FIGURE 61d

Paratyphis promontorii Stebbing, 1888 (cont.)

16 Pereiopod 5, female
17 Pereiopod 5, male
18 Pereiopod 6, female
19 Pereiopod 6, male
20 Pereiopod 7, female
21 Pereiopod 7, male
22 Uropods and telson, male

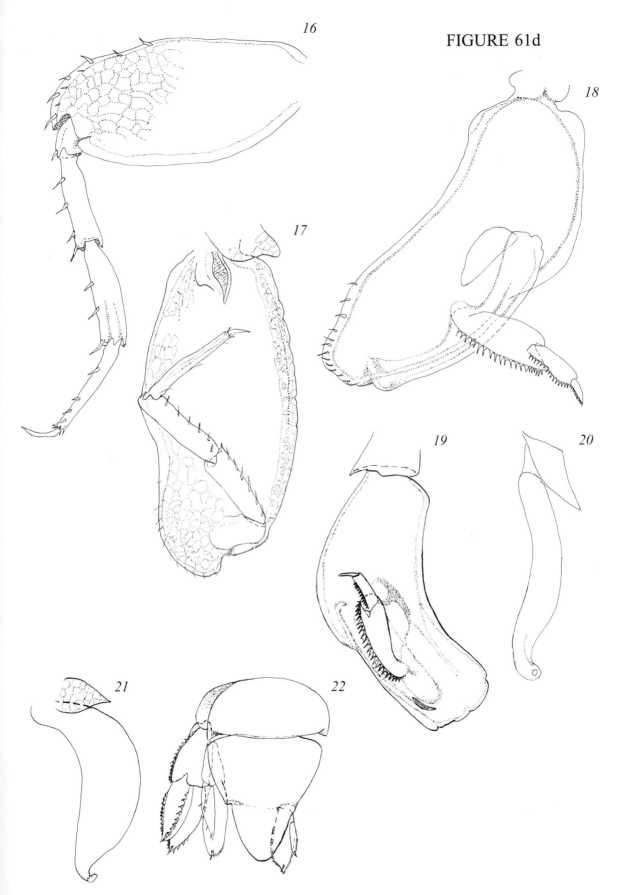

FIGURE 61d

FIGURE 62a

Tetrathyrus forcipatus Claus, 1879

Female, 3 mm; male, 4 mm

1 Lateral view, male
2 Antennula, male
3 Antenna, male
4 Pereiopod 1, male, with details
5 Pereiopod 2, male, with detail
6 Pereiopod 4, male
7 Pereiopod 5, female

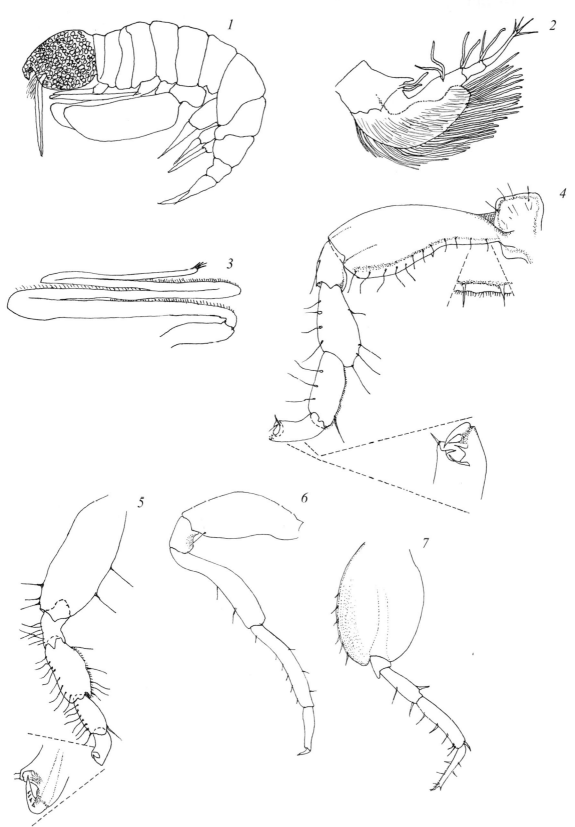

FIGURE 62b

Tetrathyrus forcipatus Claus, 1879 (cont.)

8   Pereiopod 5, male
9   Pereiopod 6, female
10 Pereiopod 6, male
11 Pereiopod 7, male
12 Uropod 1
13 Uropods and telson

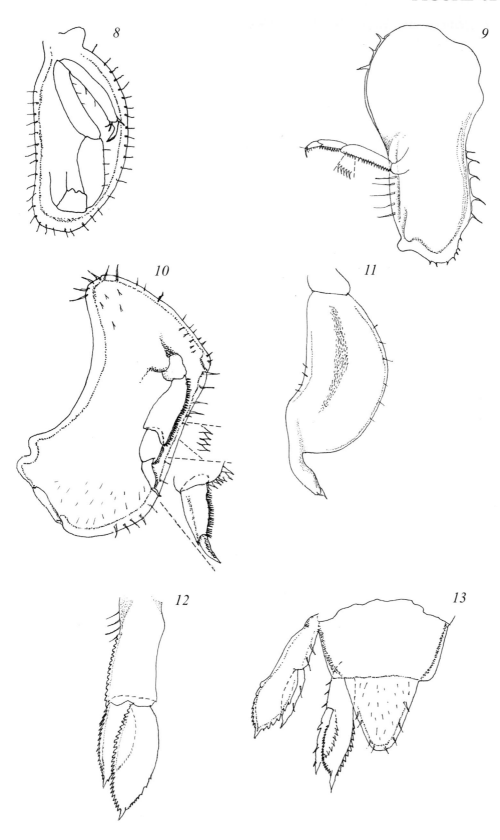

FIGURE 63a

Tetrathyrus arafurae Stebbing, 1888

Male, 3 mm

1   Lateral view
2   Antennula
3   Antenna
4   Mandibula
5   Maxillula
6   Maxilla
7   Maxilliped

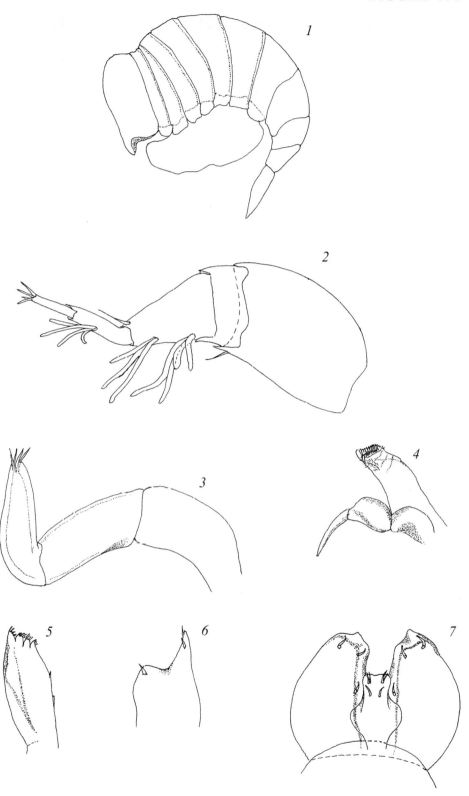

FIGURE 63b

Tetrathyrus arafurae Stebbing, 1888 (cont.)

8   Pereiopod 1
9   Pereiopod 2, with detail
10 Pereiopod 3

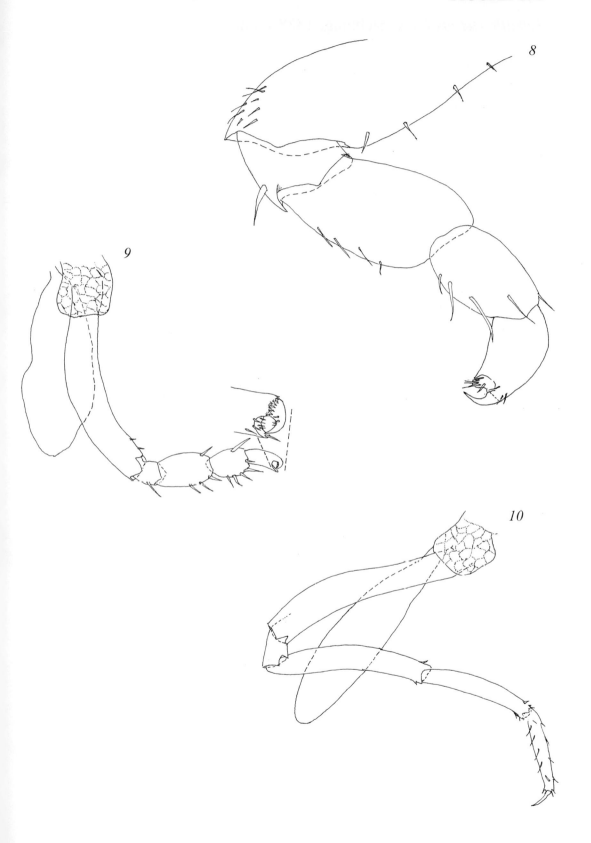

FIGURE 63c

Tetrathyrus arafurae Stebbing, 1888 (cont.)

11 Pereiopod 4
12 Pereiopod 5
13 Pereiopod 6
14 Pereiopod 7, with detail
15 Uropod 1
16 Uropods 2, 3 and telson

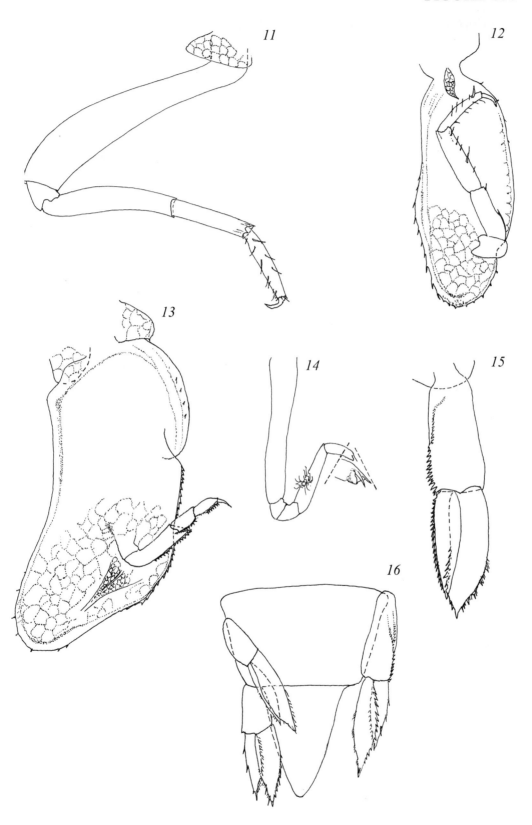

FIGURE 64a

Amphithyrus sculpturatus Claus, 1879

Male, 4.5 mm

1   Lateral view
2   Antennula, with detail
3   Antenna
4   Mandibula
5   Tip of maxillular palp
6   Maxilliped

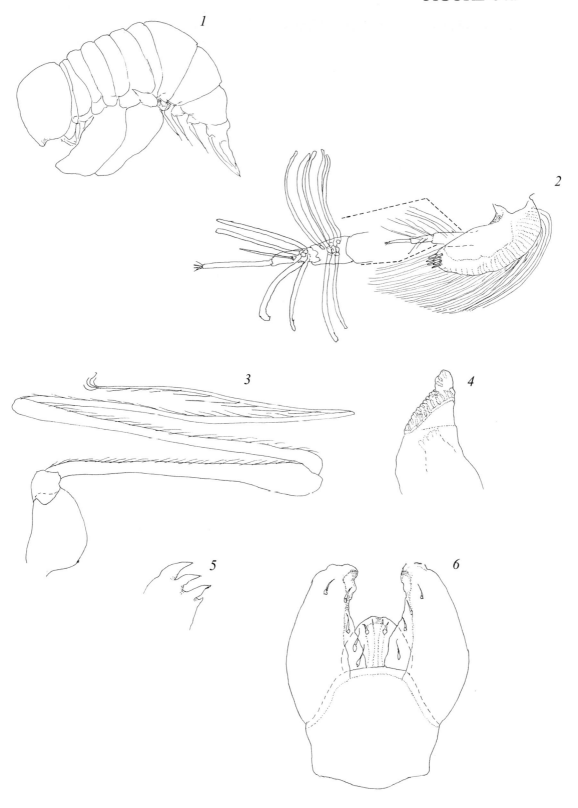

FIGURE 64b

Amphithyrus sculpturatus Claus, 1879 (cont.)

7   Pereiopod 1, left, with detail
8   Pereiopod 1, right, last segments
9   Pereiopod 2, with detail
10 Pereiopod 3
11 Pereiopod 4

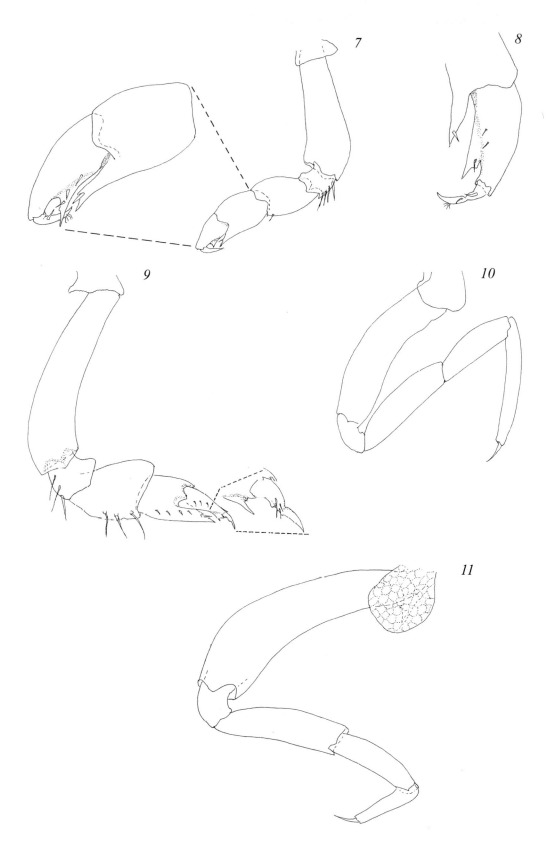

FIGURE 64c

Amphithyrus sculpturatus Claus, 1879 (cont.)

12 Pereiopod 5, with details
13 Pereiopod 6, with detail
14 Pereiopod 7, with sculptured cuticle
15 Coupling hooks and bifid spine of pleopod 1
16 Uropods and telson

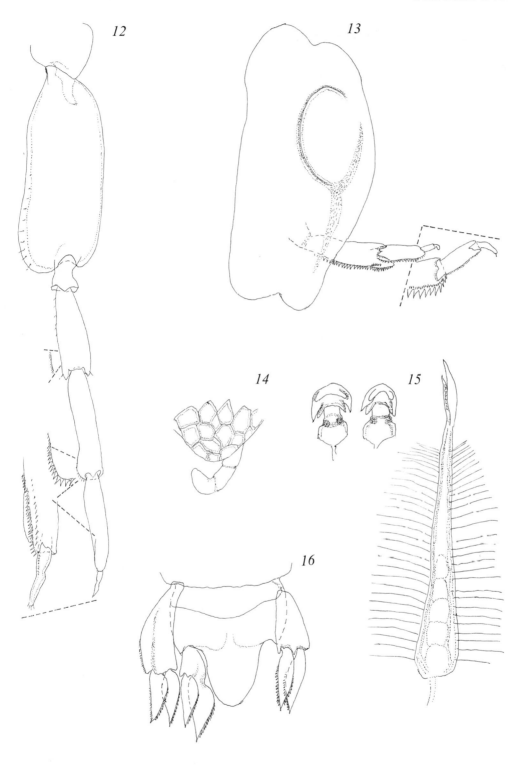

FIGURE 65a

PARASCELIDAE Claus

Parascelus typhoides Claus, 1879

Female, 4 mm; male, 7 mm

1   Lateral view, female
2   Antennula, female
3   Antenna, female
4   Mandibulae, right and left, female
5   Mandibula, male
6   Maxillular palp
7   Maxilla, female
8   Maxilliped, female
9   Maxilliped, male

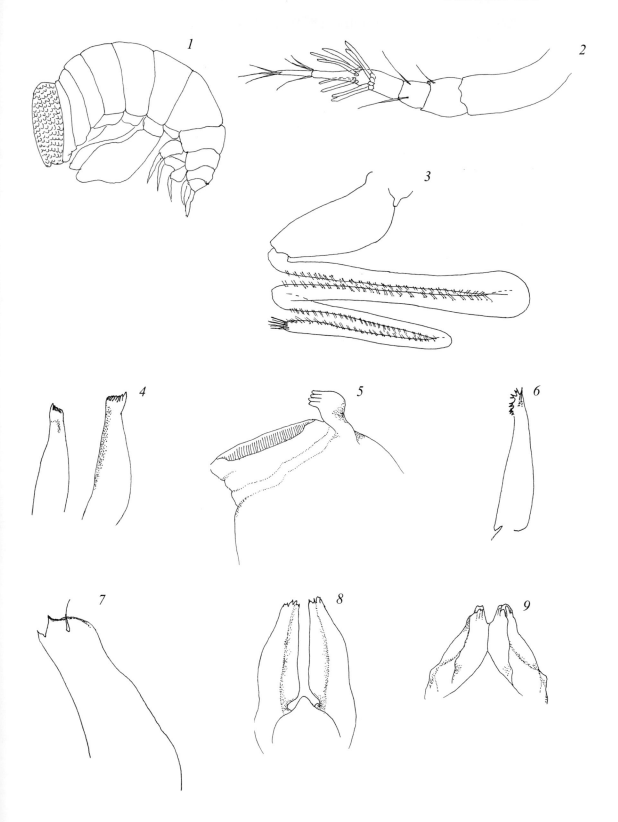

FIGURE 65b

Parascelus typhoides Claus, 1879 (cont.)

10 Pereiopod 1, female
11 Pereiopod male, with detail
12 Pereiopod 2, female
13 Pereiopod 2, male, with detail

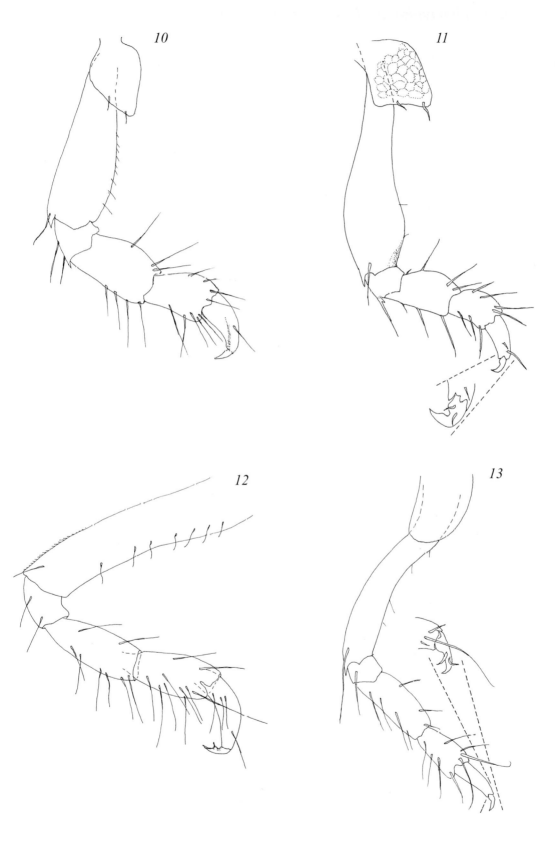

FIGURE 65c

Parascelus typhoides Claus, 1879 (cont.)

14 Pereiopod 3, female
15 Pereiopod 4, female
16 Pereiopod 5, female
17 Pereiopod 5, male

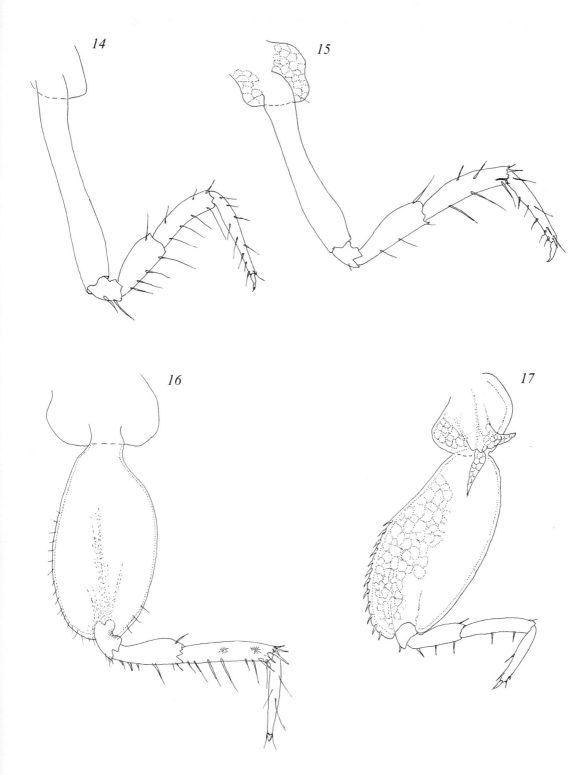

FIGURE 65d

Parascelus typhoides Claus, 1879 (cont.)

18 Pereiopod 6, female
19 Pereiopod 6, male, with detail
20 Pereiopod 7, female, with detail
21 Pereiopod 7, male
22 Uropods and telson, male, with details

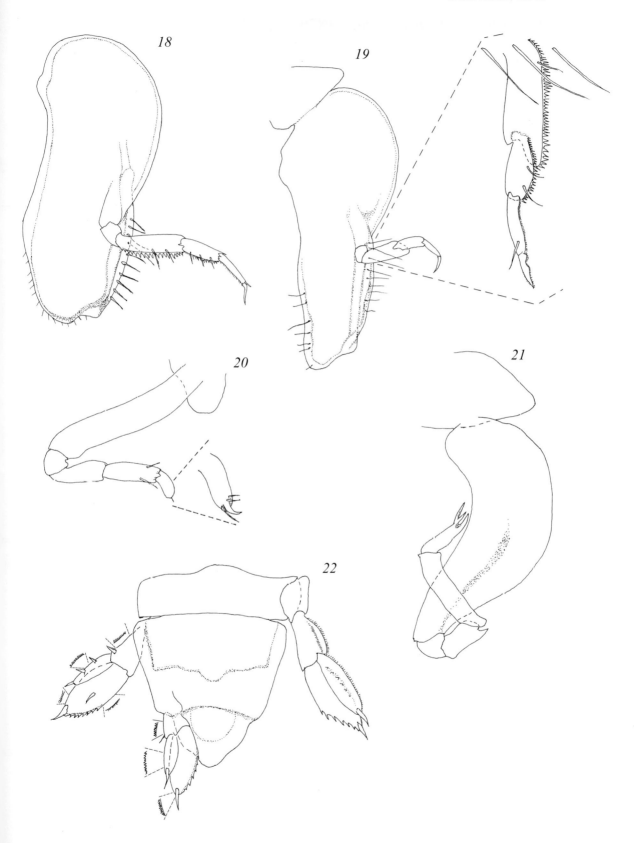

FIGURE 66a

Parascelus edwardsi Claus, 1879

Female, 3 mm; male, 4 mm

1   Female, lateral view
2   Antennula, female
3   Antennula, male
4   Antenna, female
5   Antenna, male
6   Mandibula, female
7   Mandibula, male
8   Maxillula, male
9   Maxilliped, female

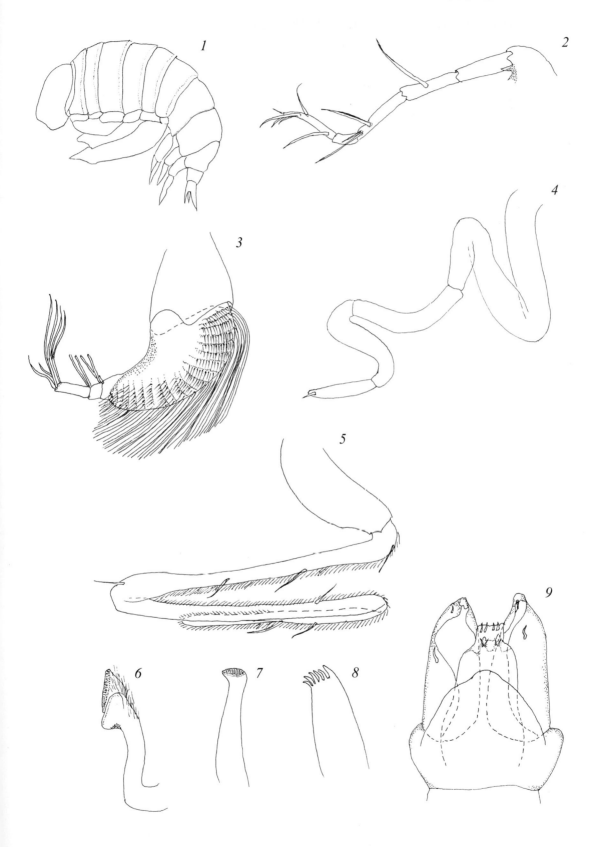

FIGURE 66b

Parascelus edwardsi Claus, 1879 (cont.)

10 Pereiopod 1, male, with detail
11 Pereiopod 2, male
12 Pereiopod 5, male
13 Pereiopod 6, female, with enlarged dactyl

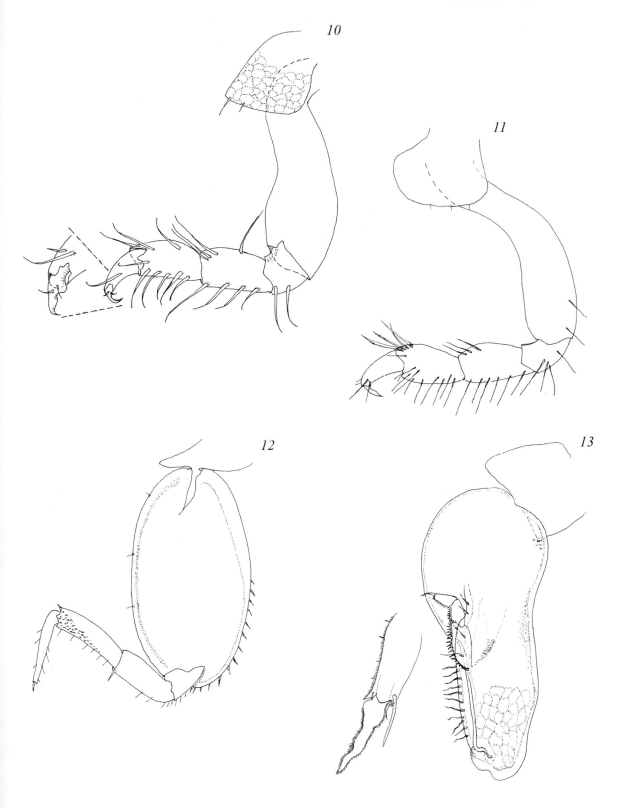

FIGURE 66c

Parascelus edwardsi Claus, 1879 (cont.)

14 Pereiopod 6, male, with details
15 Pereiopod 7, female, with detail
16 Pereiopod 7, male, with detail
17 Uropod 1
18 Uropods and telson

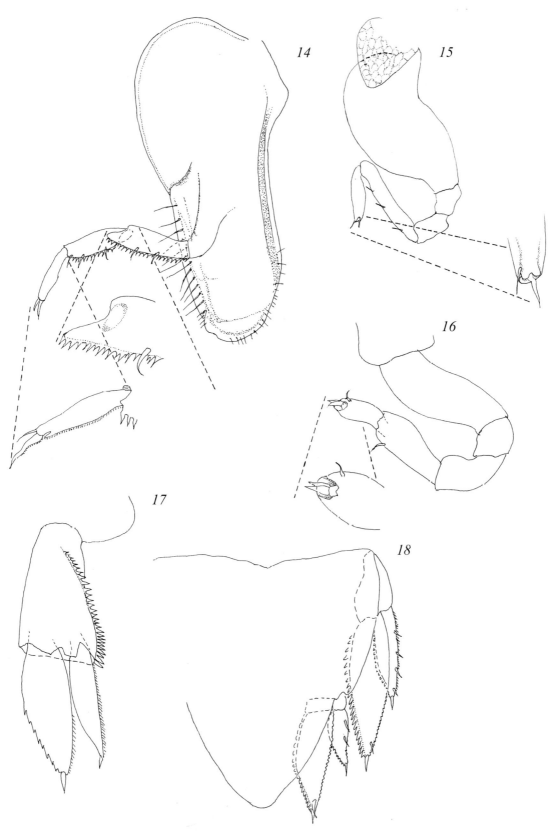

PLATES

ULTRASTRUCTURES OF MOUTHPARTS OF
PHRONIMA spp.

The morphology of the mouthparts and their ultrastructure has not been described before in the rather extensive literature on the Hyperiidea. Likewise, the anatomy of the mouthparts has not been used as a diagnostic character for the different species. One finds in the literature schematic drawings of the mouthparts complex at most, often confusingly designated (e.g., Bowman, 1973, 1978). The details of the mouthparts armature have invariably been overlooked. This publication is a follow-up to a previous one in which the ultrastructure of Hyperiidea pereiopods was revealed (Zelickman and Por, 1996).

The high magnification used here in our drawings clearly shows significant differences in the spinulation and serrulation of the mouthparts between the different species. Our few highly magnified microphotographs of the mouthparts of *Phronima* spp., made using a JEOL JSM-35 scanning electron microscope, are the first of their kind in the Hyperiidea literature. These microstructures, besides being diagnostic characters, may also give us an insight into their possible functions. The complex microstructures of the mouthparts suggest not only a function in the selecting and triturating of the food particles, but also in their digestion and absorption. The structures possibly point to a rapid processing and uptake of the soft parts of jellyfish and thaliacean tunicate prey, the empty bodies of which ultimately end up serving as a shelter for the predatory hyperiidean itself.

Key to abbreviations

Md   mandibula
Mx1   maxillula
Mx2   maxilla
Mxp   maxilliped

PLATE I

Phronima sedentaria Forskål, 1775

Females, body length 28 and 42mm

A. General view of mouthparts complex
   1   Mxp, outer lobes
   2   Mxp, inner lobe
   3   Mx2
   4   Mx1, basal part
   5   Mx1, outer lobe
   6   Mx1, great teeth of inner lobes
   7   setal row of mandibular body
   8   Md, plates at left and right
   9   Md, incisor process

B. View of maxilliped and maxilla
   1   Mxp, basal part
   2   Mxp, outer lobes
   3   Mxp, inner lobe
   4   Mx2

C. View of maxilliped
   1   Mxp, armature of basal part
   2   outer lobes, external surface
   3   outer lobe, internal surface
   4   bordering great teeth of external edge
   5   bordering thin teeth of internal edge
   6   Mx2 fragment
   7   Mxp, inner lobe
       Different setation and spinulation is seen for rough tearing of food.

D. Partial view of mouthparts, I
   1   Mxp, tip of inner lobe
   2   Mxp, tip of outer lobe
   3   Mx2
   4   Mx1, basal part
   5   Mx1, outer lobe
   6   Mx1, apical great teeth of inner lobe
   7   Md, molar plate

PLATE I

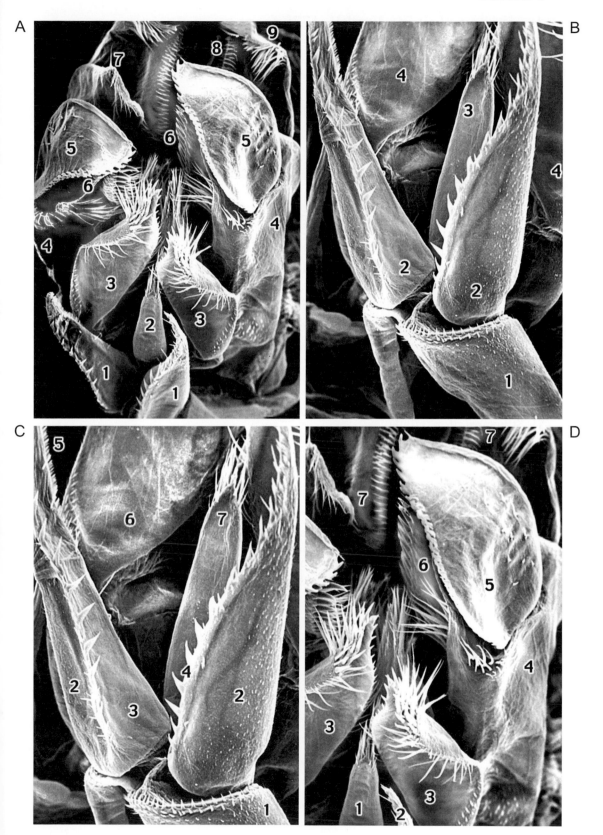

PLATE II

Phronima sedentaria Forskål, 1775 (cont.)

E. Partial view of mouthparts, II
1   Mx2
2   Mx1, outer lobe
3   Mx1, apical great teeth of inner lobe
4   Mx1, cavity in outer lobe
5   Md, molar plate
6   Md, setal row on body
7   Md, incisor process
8   fragment of maxillary duct gland

F. Partial view of mouthparts, III
1   Mx1, outer lobes
2   Mx1, apical great teeth of inner lobe
3   Md, left and right molar plate
4   setal row of mandibular body
5   Md, incisor process

G. Partial view of mandibula
1   Md, molar plate (partial)
2   Md, incisor process

H. Structures of maxillula
1a   Mx1, outer lobes
1b   internal cavities in outer lobes
2   accumulating channel of maxillary gland (?)
3   ampulla of maxillary gland (?)
4   cistern of maxillary gland (?)
5   Mx1, apex of inner lobe
6   Mx2, border
     Cavities are seen inside outer lobe since its walls are two-layered; at right, the cavity near the ampulla is half open, at left (1b), completely open.

PLATE II

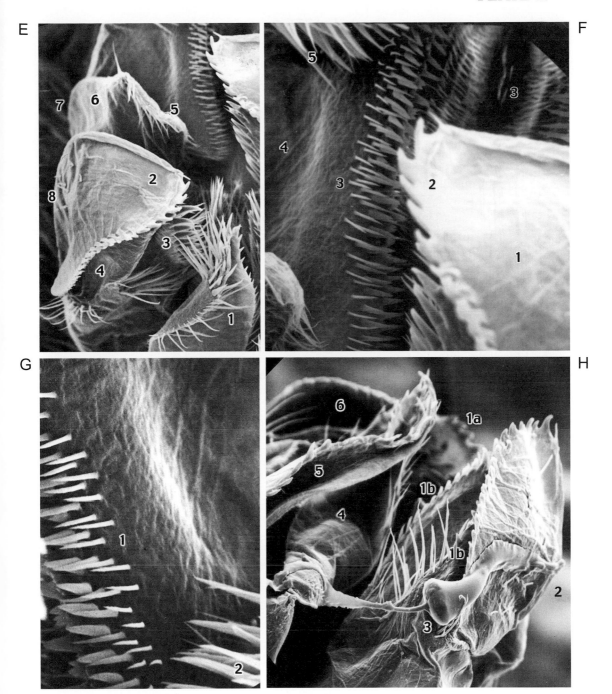

PLATE III

Phronima sedentaria Forskål, 1775 (cont.)

I. Fragments of outer and inner lobe of maxillula
   1   Collecting channel on surface of outer lobe
   2   Mx1, volucrae
       Volucrae are tapering, bifid, convoluted flat processes.
   3   semi-exposed cavity inside outer lobe
   4   Mx1, apical great teeth on inner lobe

J. View of maxillula showing great teeth
   1   Mx1, row of great teeth on apex of inner lobe
   2   Mx1, inner lobe, second row of thin spines on apex
   3   Mx1, hamate squamous micro-ornamentation on external surface of inner lobe

K. Maxillula, outer lobe
   1   Mx1, outer lobe, fragment of border with volucrae
   2   receptors

L. Maxillula, outer lobe
   1   Mx1, border with volucrae, seen on squamose surface

M. Maxillula, outer lobe
   1   Mx1, outer lobe, fragment of border with volucrae

PLATE III

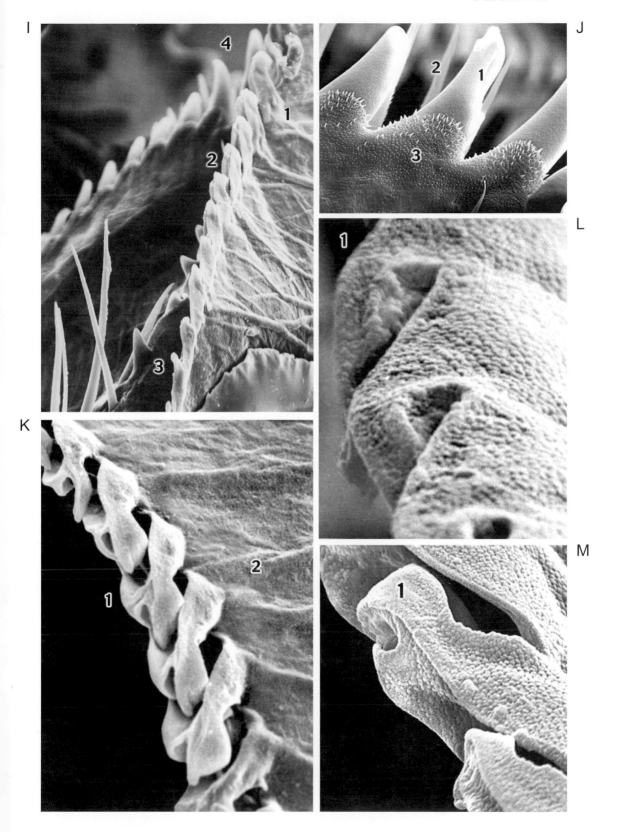

PLATE IV

Phronima sedentaria Forskål, 1775 (cont.)

N. Maxillula, outer lobe
   Mx1, outer lobe, fine structure of flat process
   Micro-ornamented surface with pores is seen.

O. Maxillula, outer lobe
   1   Mx1, surface on basal part of outer lobe near gland cone
        receptors

PLATE IV

N

O

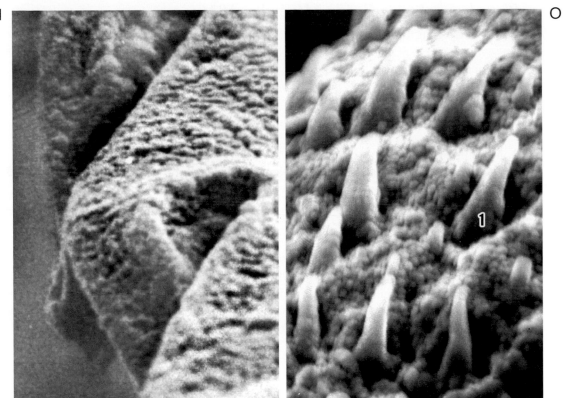

PLATE V

Phronima solitaria Guérin-Ménéville, 1836

Male, body length 44 mm

A. General view of maxilliped
    1   Mxp, outer lobes
    2   Mxp, inner lobe
    3   Mxp, basal part

B. Maxilliped, outer lobe
    1   Mxp, outer lobe (lateral–oblique angle), with thin bristles
        on internal border
    2   conical teeth on external border
    3   petaloid teeth

C. Maxilliped, inner lobe in ventro-frontal position
    1   micro-ornamented surface and differentiated type of setation:
        small bristles and great acute styliform spines with spiral
        secondary spinulation
    2   basal part of great spine, not arranged in pit, with basal heel

D. Maxillula, outer lobe
    1   apical great teeth
    2   row of conical teeth on external border

PLATE V

PLATE VI

Phronima solitaria Guérin-Ménéville, 1836 (cont.)

E. Maxillula, outer lobe
  1 apical great teeth on outer lobe
  2 accessory small triangular teeth (analogous to *Phronima atlantica,*
     Plate VIII, E:2)
  3 row of crescent shaped teeth on lateral border of outer lobe

F. Maxillula, outer lobe
  1 apex with great solid teeth
  2, 3 lateral row of crescent denticles
     On tip of denticle, probably single terminal receptor.
  4 rows of bristles
  5 thin bifurcate bristles

G. Maxillula, outer lobe
  1 crescent tooth with single terminal receptor
  2 accessory superficial secondary denticle

H. Maxilla, apical spines on lateral lobe
  1 apical spines with badge form of receptor
  2 scale-type of micro-ornamentation with hamate hooks

PLATE VI

PLATE VII

Phronima atlantica Guérin-Ménéville, 1836
Female, body length 15 mm

A. Maxillula, general view
   1   Mx1, outer lobe
   2   great teeth of inner lobe
   3   volucrae on the border
   4   Mx2, posterior

B. Maxillula, interior face, apex of outer lobe

C. Maxillula, exterior face, apex of outer lobe

D. Maxillula, upper border of outer lobe
   Lobe appears bilaminated.
   1   row of teeth
   2   row of triangular setose polyseriate processes

PLATE VII

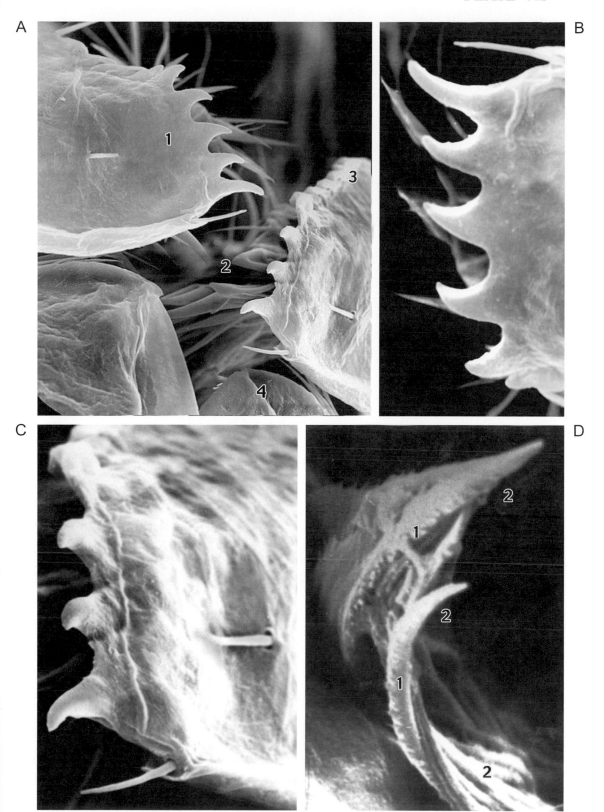

PLATE VIII

Phronima atlantica Guérin-Ménéville, 1836 (cont.)

E. Maxillula, upper border
   1   great tooth
   2   small triangular humps
   3   broken polyseriate process with secondary setal formation

F. Maxillula, polyseriate processes on surface of outer lobe
   (Part of E at higher magnification)

G. Outer lobe of maxillula, showing duct of maxillary gland (?)
   on background of outer lobe of Mx1

H. Maxilla
   1   outer lobe
   2   inner lobe

PLATE VIII

COMPREHENSIVE BIBLIOGRAPHY
OF THE HYPERIIDEA

Agraval, V. P. 1967. The feeding habit and the digestive system of *Hyperia galba*. *Proceedings of the Symposium on Crustacea held at Ernakulam. Marine Biological Association of India*, 2:545–548.

Altner, H. & Prillinger, L. 1980. Ultrastructure of invertebrate chemo-, thermo- and hygro-receptors and its functional significance. *International Revue of Cytology*, 67:69–139.

Ball, E. E. 1977. Fine structure of the compound eyes of the midwater amphipod *Phronima* in relation to behavior and habitat. *Tissue and Cell*, 9:521–536.

Barnard, K. H. 1930. Crustacea, Part XI: Amphipoda. In: *British Antarctic "Terra Nova" Expedition, 1910. Natural History Reports, Zoology*, 8(4):307–454.

— 1931. Amphipoda. In: *Great Barrier Reef Expedition, 1928–1929. Scientific Reports*, 4(4):111–135.

— 1932. Amphipoda. In: *Discovery Reports*, 5:1–326.

— 1937. Amphipoda Hyperiidea. *Australasian Antarctic Expedition, 1911–1914. Scientific Reports, Series C (Zoology and Botany)*, 2(5):1–4.

— 1937. Amphipoda. In: *The John Murray Expedition, 1933–1934. Scientific Reports*, 4(6):131–201.

Barkhatov, J. A. & Vinogradov, M. E. 1988. Amphipod — hyperiids of the subantarctic and pelagic zones of the central part of the Pacific Ocean. In: Vinogradov M. E. & Flint M. V. (eds.), *Ecosistemy subantarcticheskoi zony Tikhogo okeana*. Moscow: Nauka, pp. 228–245. [In Russian]

Bate, C. S. 1861. On the morphology of some Amphipoda of the division Hyperina. *Annals and Magazine of Natural History, Series 3*, 8(43):1–16.

— 1862. *Catalogue of the specimens of Amphipodous Crustacea in the collection of the British Museum*. London: British Museum of Natural History, 399 pp.

Behning, A. L. 1912. Die systematische Zusammensetzung und geographische Verbreitung der Familie Vibiliidae. *Zoologica (Stuttgart)*, 26(8):211–226.

— 1913. Die Vibiliiden (Amphipoda Hyperiidea) der Deutschen Südpolar-, Schwedischen Südpolar- "Albatross"- und "Michael Sars"- Expeditionen. *Zoologischer Anzeiger*, 41(12):529–534.

— 1925. Amphipoda der Deutschen Tiefsee-Expedition. I. Hyperiidea fam. Vibiliidae Claus 1872. *Wissenschaftliche Ergebnisse der Deutschen Tiefsee-Expedition auf dem Dampfer "Valdivia" 1898–1899*, 19(9):479–500.

— 1927. Die Vibiliiden der Deutschen Südpolar-Expedition 1901–1903. *Wissenschafltliche Ergebnisse der Deutschen Südpolar-Expedition*, 19(5):113–121.

— 1939. Die Amphipoda–Hyperiidea der den Fernen Osten der USSR umgrenzenden Meere. *Internationale Revue der gesamten Hydrobiologie und Hydrographie*, 38(3/4):353–367.

— & Woltereck, R. 1912. Achte Mitteilung über die Hyperiiden der "Valdivia-" Expedition insbesondere über die Vibiliiden. *Zoologischer Anzeiger,* 41(1):1–11.

Bonnier, T. 1896. Résultats scientifiques de la campagne du "Caudan" dans le golfe de Gascogne. Edriophthalmes. *Annales de l'Université de Lyon,* 26:527–689.

Bouchon, D. & Chaigneau, J. 1991. Comparison of cuticular adhesive structures linking anatomical parts in Crustacea, and their adaptive significance (Decapoda and Isopoda). *Crustaceana,* 60(1):7–17.

Bousfield, E. L. 1951. Pelagic Amphipoda of the Belle Isle Strait Region. *Journal of the Fisheries Research Board of Canada,* 8(3):134–163.

— 1956. Studies on the Shore Crustaceans collected in Eastern Nova Scotia and Newfoundland, 1954. *Bulletin of the National Museum of Canada,* 142:127–152.

Bovallius, C. 1887. Contributions to a monograph of the Amphipoda Hyperiidea. Part I. The families Tyronidae, Lanceolidae and Vibiliidae. *Kongliga Svenska Vetenskapsakademiens Handlingar,* 21(5):1–72.

— 1889. Contributions to a monograph of the Amphipoda Hyperiidea. Part I:2. The families Cyllopodidae, Paraphronimidae, Thaumatopsidae, Mimonectidae, Hyperiidae, Phronimidae and Anchylomeridae. *Kongliga Svenska Vetenskapsakademiens Handlingar,* 22(7):1–434.

— 1890. The Oxycephalids. *Nova Acta Regiae Societatis Scientiarum. Uppsala, Series 3,* 15:1–141.

Bowman, T. E. 1953. *The Systematics and Distribution of pelagic Amphipods of the Families Vibiliidae, Paraphronimidae, Hyperiidae, Dairellidae and Phrosinidae from the Northeastern Pacific,* Ph. D. Thesis, University of California, Los Angeles, 480 pp.

— 1960. The Pelagic Amphipod genus *Parathemisto* (Hyperiidea: Hyperiidae) in the North Pacific and Adjacent Arctic Ocean. *Proceedings of the United States National Museum,* 112(3439):343–392.

— 1973. Pelagic amphipods of the genus *Hyperia* and closely related genera (Hyperiidea: Hyperiidae). *Smithsonian Contributions to Zoology,* 136:1–76.

— 1978. Revision on the pelagic amphipod genus *Primno* (Hyperiidea: Phrosinidae). *Smithsonian Contributions to Zoology,* 275:1–23.

— 1985. The correct identity of the pelagic amphipod *Primno macropa,* with a diagnosis of *Primno abyssalis* (Hyperiidae: Phrosinidae). *Proceedings of the Biological Society of Washington,* 98(1):121–126.

— & Abele, L. G. 1982. Classification of the recent Crustacea. In: D. E. Bliss (ed.), *The Biology of Crustacea,* New York: Academic Press, Vol. 1, pp. 1–27.

—, Cohen, A. C. & McGuiness M. M. 1982. Vertical distribution of *Themisto gaudichaudii* (Amphipoda: Hyperiidea) in Deepwater Dumpsite 106 off the mouth of Delaware Bay. *Smithsonian Contributions to Zoology,* 351:1–24.

— & Gruner, H.-E. 1973. The families and genera of Hyperiidea (Crustacea: Amphipoda). *Smithsonian Contributions to Zoology,* 146:1–64.

—, Meyers, C. D. & Hicks, S. D. 1963. Notes on association between hyperiid amphipods and medusae in Chesapeake and Narragansett Bays and the Niantic River. *Chesapeake Science,* 4(3):141–146.

Brusca, G. J. 1967. The ecology of pelagic Amphipoda. II. Observations on the reproductive cycles of several pelagic amphipods from the waters off southern California. *Pacific Science*, 21(4):449–456.

— 1973. Pelagic amphipods from the waters near Oahu, Hawaii, excluding the family Scinidae. *Pacific Science*, 27(1):8–27.

— 1978. Contribution to the knowledge of hyperiid amphipods of the family Scinidae from near Hawaii, with a description of a new species, *Scina hawaiensis*. *Pacific Science*, 32(3):280–292.

— 1981. Annotated keys to the Hyperiidea (Crustacea: Amphipoda) of North American coastal waters. *Technical Report of the Allan Hancock Foundation*, 5:1–76.

— 1981. On the anatomy of *Cystisoma* (Amphipoda: Hyperiidea). *Journal of Crustacean Biology*, 1:358–375.

Cecchini, C. 1929. Oxicefalidi del Mar Rosso. *Annali Idrografici Genova Memorii*, 16 pp.

Chevreux, E. 1900. Amphipodes Provenant des Campagnes de l'"Hirondelle" (1885–1888). In: *Résultats des Campagnes Scientifiques Accomplies sur son Yacht par Albert Ier, Prince Souverain de Monaco*, 16:1–195.

— 1935. Amphipodes Provenant des Campagnes Scientifiques du Prince Albert Ier de Monaco. In: *Résultats des Campagnes Scientifiques Accomplies sur son Yacht par Albert Ier, Prince Souverain de Monaco*, 90:1–214.

— & Fage, L. 1925. *Amphipodes. Faune de France*, 9:1–488.

Claus, C. 1879. Der Organismus der Phronimiden. *Arbeiten aus den Zoologischen Instituten der Universität zu Wien*, 2:59–146.

— 1879. Die Gattungen und Arten der Platysceliden in systematischer Übersicht. *Arbeiten aus den Zoologischen Instituten der Universität zu Wien*, 2:5–43, 147–198.

— 1887. *Die Platysceliden*. Wien: Alfred Hölder, 77 pp.

Coleman, C. O. 1992. Foregut morphology of Amphipoda (Crustacea). An example of its relevance for systematics. *Ophelia*, 36:135–150.

— 1994. Comparative anatomy of the alimentary canal of hyperiid amphipods. *Journal of Crustacean Biology*, 14(2):346–370.

Colosi, G. 1918. Oxicefalidi: Raccolte planctoniche fatte dalla R. Nave "Liguria" 1903–1905, II, fasc. 3. Crostacei. *Pubblicazioni del Real Istituto di Studii Superiori Firenze*, pp. 207–224.

Dahl, E. 1959. The amphipod, *Hyperia galba*, an ecto-parasite of the jelly-fish *Cyanea capillata*. *Nature*, 183(4677):1749.

— 1977. The amphipod functional model and its bearing upon systematics and phylogeny. *Zoologica Scripta*, 6:221–228.

— 1979. Deep-sea carrion feeding amphipods: evolutionary patterns in niche adaptation. *Oikos*, 33:167–175.

Dakin, W. J. & Colefax, A. N. 1940. The plankton of the Australian coastal waters off New South Wales. 1. *Publications of the University of Sydney, Department of Zoology*, 1:1–215.

Daniel, R. 1973. Siphonophores and their commensals in the Indian Ocean. *Journal of the Marine Biological Association of India*, 15(1):354–358.

Dick, R. I. 1970. Hyperiidea (Crustacea: Amphipoda): Keys to South African genera and species and a distribution list. *Annals of the South African Museum*, 57(3):25–86.

Diebel, C. E. 1988. Observations on the anatomy and behaviour of *Phronima sedentaria* (Forskål) (Amphipoda: Hyperiidea). *Journal of Crustacean Biology*, 8(1):79–90.

—1992. Arrangement and external morphology of sensilla on the dorsal surface of three genera of hyperiid amphipods (*Phronima, Lycaea,* and *Vibilia*). *Journal of Crustacean Biology*, 12(4):714–728.

Dittrich, B. U. 1986. *Beiträge zur Biologie und Ökologie von Hyperia galba (Montagu, 1813)*. Dissertation, Universität Bochum, 202 pp.

—1987. Postembryonic development of the parasitic amphipod *Hyperia galba*. *Helgoländer Wissenschaftliche Meeresuntersuchungen*, 41:217–232.

—1988. Studies on the life cycle and reproduction of the parasitic amphipod *Hyperia galba* in the North Sea. *Helgoländer Wissenschaftliche Meeresuntersuchungen*, 42:79–98.

—1992. Functional morphology of the mouthparts and feeding strategies of the parasitic amphipod *Hyperia galba* (Montagu, 1813). *Sarsia*, 77:11–18.

Dudich, E. 1926. Systematische und Biologische Studien an den *Phronima* — Arten des Golfes von Neapel. *Zoologischer Anzeiger*, 65:117–139.

Dunbar, M. J. 1954. The amphipod Crustacea of Ungava Bay, Canadian Eastern Arctic. *Journal of the Fisheries Research Board of Canada*, 11(6):709–798.

—1957. The determinants of production in Northern Seas: A study of the biology of *Themisto libellula* Mandt. *Canadian Journal of Zoology*, 35:797–819.

—1963. Amphipoda, Sub-order: Hyperiidea. Family: Hyperiidae. *Zooplankton Sheet*, Conseil International pour l'Exploration de la Mer, 103:1–4.

Echelman, T. 1989. *The seasonal surface near reef and offshore zooplankton dynamics near Eilat (Gulf of Aqaba), Red Sea.* M.Sc. Thesis, Tel-Aviv University.

Evans, F. 1968. The subgenera *Parathemisto* and *Euthemisto* of the genus *Parathemisto* (Amphipoda, Hyperiidea). *Crustaceana*, 14:105–106.

—& Sheader, M. 1972. Host species of the hyperiid amphipod *Hyperoche medusarum* (Krøyer) in the North Sea. *Crustaceana*, supplement 3:275–276.

Fage, L. 1954. Les Amphipodes pélagiques du genre *Rhabdosoma*. *Comptes Rendus des Séances Hebdomadaires de l'Académie des Sciences* (Paris), 239(11):661–663.

—1960. Oxycephalidae (Amphipodes Pélagiques). *"Dana" Report*, 9(52):1–145.

Flores, M. & Brusca, G. J. 1975. Observations on two species of hyperiid amphipods associated with the ctenophore *Pleurobrachia bachei*. *Bulletin of the Southern California Academy of Sciences*, 74:10–15.

Franqueville, C. 1971. Macroplancton profond (Invertébrés) de la Mediterranée Nord-occidentale. *Téthys*, 3(1):11–56.

Fukuchi, U. 1977. Regional distribution of Amphipoda and Euphausiacea in the northern North Pacific and Bering Sea in Summer of 1969. *Research Institute of North Pacific Fisheries*, Special Volume, pp. 439–459.

Fulton, J. 1968. A laboratory manual for the identification of British Columbia marine zooplankton. *Fisheries Research Board of Canada Technical Report*, 55:1–141.

Gamulin, T. 1948. Prilog Pozanavanju Zooplanktona Srednjedalmatinskog Otocnog Podrucja. *Acta Adriatica*, 3(7):159–194.

Garbowsky, T. 1896. Hyperiden-artige Amphipoden des Mittelmeeres. Monographie bearbeitet auf Grund des während der 5. Expedition S. M. Schiffes "Pola" (1890–94)

[430]

gesammelten Materials. 1. Die Sciniden. *Denkschriften der Kaiserlichen Akademie der Wissenschaften in Wien, Mathematisch- naturwissenschaftliche Klasse*, 63:1–89.

Giles, G. M. 1887. Natural history notes from Her Majesty Indian marine survey steamer "Investigator". On six new amphipods from the Bay of Bengal. *Journal of the Asiatic Society of Bengal*, 56, part 2(2):212–229.

Gray, J. S. & McHardy, R. A. 1967. Swarming of hyperiid amphipods. *Nature*, 215(5096):100.

Hallberg, E., Nillsson, H. L. & Elofsson, R. 1980. Classification of amphipod compound eyes — the fine structure of the ommatidial units (Crustacea, Amphipoda). *Zoomorphologie*, 94:279–306.

Harbison, G. R. 1976. The development of *Lycaea pulex* Marion, 1874 and *Lycaea vincentii* Stebbing 1888 (Amphipoda, Hyperiidea). *Bulletin of Marine Science*, 26:152–164.

—, Biggs, D. C. & Madin, L. P. 1977. The associations of Amphipoda Hyperiidea with gelatinous zooplankton. II. Associations with Cnidaria, Ctenophora and Radiolaria. *Deep-Sea Research*, 24:465–488.

Hardy, A. C. & Gunter, E. R. 1935. The plankton of the South Georgia Whaling Grounds and adjacent waters, 1926–1927. *Discovery Reports*, 11:1–456.

Hollowday, E. D. 1947. On the commensal relationship between the amphipod *Hyperia galba* (Mont.) and the scyphomedusa *Rhizostoma pulmo* Agassiz, var. *octopus* Oken. *Journal of the Quekett Microscopical Club, Series 4*, 2(4):187–190.

Holmes, S. J. 1908. The Amphipoda collected by U. S. Bureau of Fisheries Steamer "Albatross" off the west coast of North America in 1903 and 1904, with descriptions of a new family and several new genera and species. *Proceedings of the United States National Museum*, 35(1654):489–543.

Hure, J. 1955. Distribution annuelle verticale du zooplancton sur une station de l'Adriatique Méridionale. *Acta Adriatica*, 7(7):1–72.

— 1961. Dnevna Migracija i Sezonska Vertikalna Raspodjela Zooplanktona Dubljeg Mora. *Acta Adriatica*, 9(6):1–59.

—, di Carlo, B. S. & Basile, A. 1971. Comparazione tra lo zooplancton del Golfo di Napoli e dell'Adriatico Meridionale presso Dubrovnik. II. Amphipoda (Hyperiidea). *Pubblicazioni della Stazione Zoologica di Napoli*, 37:599–609.

Hurley, D. E. 1955. Pelagic amphipods of the suborder Hyperiidea in New Zealand waters. I. Systematics. *Transactions of the Royal Society of New Zealand*, 83(1):119–194.

— 1956. Bathypelagic and other Hyperiidea from Californian waters. *Allan Hancock Foundation Publications, Occasional Paper*, 18:1–25.

— 1960. Amphipoda Hyperiidea. *B.A.N.Z. Antarctic Research Expedition 1929–1931 Reports, Series B (Zoology and Botany)*, 8(5):107–113.

Ikeda, T. 1990. A growth model of a hyperiid amphipod *Themisto japonica* (Bovallius) in the Japan Sea, based in its intermoult period and moult increment. *Journal of the Oceanographic Society of Japan*, 46(6):261–271.

— 1991. Assimilated carbon budget for the hyperiid amphipod *Themisto japonica* (Bovallius) from the Japan Sea as influenced by temperature. *Journal of the Oceanographic Society of Japan*, 47(1):7–16.

— 1995. Distribution, growth and life cycle of the mesopelagic amphipod *Primno abyssalis* (Hyperiidea: Phrosinidae) in the southern Japan Sea. *Marine Biology*, 123:789–798.

—, Hirakawa, K. & Imamura, A. 1992. Abundance, population structure and life cycle of a hyperiid amphipod *Themisto japonica* (Bovallius) in Toyama Bay, southern Japan Sea. *Bulletin of the Plankton Society of Japan*, 39:1–16.

Jacques, F. 1989. The setal system of crustaceans: types of setae, groupings, and functional morphology. In: B. E. Felgenhauer, L. Watling & A. B. Th tle (eds.), *Functional morphology of feeding and grooming in Crustacea*. Rotterdam: A. A. Balkema, pp. 1–13.

Kane, J. E. 1962. Amphipoda from waters south of New Zealand. *New Zealand Journal of Science*, 5(3):295–315.

—1963. Stages in the early development of *Parathemisto gaudichaudii* (Guér.) (Crustacea Amphipoda: Hyperiidea). The development of secondary sexual characters and of the ovary. *Transactions of the Royal Society of New Zealand, Zoology*, 3(5):35–45.

Kim, C. B. & Kim, W. 1993. Phylogenetic relationships among gammaridean families and amphipod suborders. *Journal of Natural History*, 27:933–946.

Land, M. F. 1981. Optics of the eyes of *Phronima* and other deep-sea amphipods. *Journal of Comparative Physiology*, 145:209–226.

—1989. The eyes of hyperiid amphipods: relations of optical structure to depth. *Journal of Comparative Physiology*, 146(6):751–762.

—1992. Locomotion and visual behaviour of mid-water crustaceans. *Journal of the Marine Biological Association of the United Kingdom*, 72:41–60.

Laval, Ph. 1963. Sur la biologie et les larves de *Vibilia armata* Bov. et de *V. propinqua* Stebb., Amphipodes Hypérides. *Comptes Rendus Hebdomadaires des Séances de l'Académie des Sciences* (Paris), 257(6):1389–1392.

—1966. *Bougisia ornata*, genre et espèce nouveaux de la famille des Hyperiidae (Amphipoda: Hyperiidea). *Crustaceana*, 10:210–218.

—1968. Observations sur la biologie de *Phronima curvipes* Voss. (Amphipode Hypéride) et description du mâle adulte. *Cahiers de Biologie Marine*, 9:347–362.

—1968. Développement en élevage et systématique d'*Hyperia schizogeneios* Stebb. (Amphipode, Hypéride). *Archives de Zoologie expérimentale et générale*, 109(1):25–67.

—1970. Sur des Phronimidae de l'Océan Indien et de l'Océan Pacifique, avec la validation de *Phronima bucephala* Giles, 1887 comme espèce distincte de *P. colletti* Bov., 1887 (Crustacés Amphipodes). *Cahiers O.R.S.T.O.M., série Océanographique*, 8(1):47–57.

—1972. Comportement, parasitisme et écologie d'*Hyperia schizogeneios* Stebb. (Amphipode Hypéride) dans le plancton de Villefranche-sur-Mer. *Annales de l'Institut Océanographique, Paris*, 48(1):49–74.

—1975. Une analyse multivariable du développement au laboratoire de *Phronima sedentaria* (Forsk.), Amphipode hypéride. *Annales de l'Institut Océanographique, Paris*, 51(1):5–41.

—1978. The barrel of the pelagic amphipod *Phronima sedentaria* (Forsk.) (Crustacea: Hyperiidea). *Journal of Experimental Marine Biology and Ecology*, 33:187–211.

—1979. *Contribution à l'étude des amphipodes hypérides*. *Dissertation Abstr. Int.* (c) 39(4):1–650.

—1980. Hyperiid amphipods as crustacean parasitoids associated with gelatinous zooplankton. *Oceanography and Marine Biology*, 18:11–56.

—1981. Relations entre la femelle et le tonnelet chez *Phronima sedentaria* (Forsk.),

(Amphipode hypéride). *Rapports et Procés-Verbaux des Réunions de la Commission internationale pour l'Exploration scientifique de la mer Méditerranée*, 27(7):193–194.

—, Braconnot, J.-C., Carré, C., Goy, J., Morand, P. & Mills C. E., 1989. Small-scale distribution of macroplankton and micronekton in the Ligurian Sea (Mediterranean Sea) as observed from the manned submersible "Cyana". *Journal of Plankton Research*, 11(4):665–685.

Lewis, J. B. & Fish, A. G. 1969. Seasonal variation of the zooplankton fauna of surface waters entering the Caribbean Sea at Barbados. *Caribbean Journal of Sciences*, 9(1–2):1–24.

Lo Bianco, S. 1902. Le pesche pelagiche abissali eseguite da F. A. Krupp col yacht "Puritan" nelle adiacenze di Capri. *Mitteilungen aus der Zoologischen Station Neapel*, 15(3):413–482.

Lorz, H. & Pearcy, W. G., 1975. Distribution of hyperiid amphipods off the Oregon Coast. *Journal of the Fisheries Research Board of Canada*, 32(8):1442–1447.

Mackie, G. O., Pugh, P. R. & Purcell, J. E. 1987. Siphonophore biology. *Advances in Marine Biology*, 24:97–262.

Madin, L. P. & Harbison, G. R. 1977. The associations of Amphipoda Hyperiidea with gelatinous zooplankton. 1. Associations with Salpidae. *Deep-Sea Research*, 24(5):449–463.

Mauchline, T. & Ballantine, A. R. S. 1975. The integumental organs of Amphipods. *Journal of the Marine Biological Association of the United Kingdom*, 55:345–355.

McLaughlin, P. A. 1980. *Comparative morphology of recent Crustacea*, San Francisco: W. H. Freeman & Company, 177 pp.

—1982. Comparative morphology of crustacean appendages. In: Bliss, D. E. (ed.), *The Biology of Crustacea*, New York: Academic Press, Vol. 2, pp. 197–256.

Meruane, Z. 1982–1983. Anfipodos hyperidos recolectados en los aguas circundantes a las islas Robinson Crusoe y Santa Clara. *Investigationes marines*, 10(1–2):35–40.

Meyer-Rochow, V. B. 1978. The eyes of mesopelagic crustaceans. II. *Streetsia challengeri* (Amphipoda). *Cell and Tissue Research*, 186:337–349.

Minchin, D. & Holmes, J. M. C. 1987. *Phronima sedentaria* (Forskål) (Crustacea: Amphipoda: Hyperiidea) in Irish waters. *Irish Natural History Journal*, 22(5):202–203.

Minkiewicz, R. 1909. Mémoire sur la biologie du tonnelier de mer (*Phronima sedentaria* Forsk.). *Bulletin de l'Institut Océanographique de Paris*, 152:1–21.

Mogk, H. 1926. Versuch einer Formanalyse bei Hyperiden. *International Revue für Hydrobiologie und Hydrographie*, 14:160–192.

Möller, H. 1978–1979. Significance of coelenterates in relation to other plankton organisms. *Meeresforschung*, 27:1–18.

Nayar, K. N. 1959. The Amphipoda of the Madras coast. *Bulletin of the Madras Government Museum, N. S.* (Natural History Section), 6(3):1–59.

Nillsson, D.-E. 1982. The transparent compound eye of *Hyperia* (Crustacea): examination with a new method for analysis of refractive index gradients. *Journal of Comparative Physiology*, 147:339–349.

Pillai, N. K. 1966. Pelagic amphipods in the collections of the Central Marine Fisheries

Research Institute, India. Parts I–II. In: *Proceedings of the Symposium on Crustacea held at Ernakulam. Marine Biological Association of India*, pp. 169–232.

Pirlot, J. M. 1929. Résultats zoologiques de la croisière Atlantique de l'"Armauer Hansen" (1922). I. Les Amphipodes Hypérides. *Mémoires de la Société Royale des Sciences de Liège, Series 3*, Volume 15(2):1–196.

— 1930. Les Amphipodes de l'expédition du "Siboga". Part I: Les Amphipodes Hypérides (à l'exception des Thaumatopsidae et des Oxycephalidae). *Siboga-Expedition, Reports*, 33a:1–55.

— 1932. Introduction à l'étude des amphipodes hypérides. *Annales de l'Institut Océanographique de Paris*, N.S., 12(1):1–36.

— 1938. Amphipodes de l'Expédition du "Siboga". 2. Les Amphipodes littoraux. Addendum: Les Amphipodes hypérides. *Siboga-Expédition, Monographie*, 33f:329–388.

— 1939. Sur des Amphipodes Hypérides provenant des croisières du Prince Albert Ier de Monaco. In: *Résultats des Campagnes Scientifiques Accomplies sur son Yacht par Albert Ier, Prince Souverin de Monaco*, 102:1–63.

Reid, D. M. 1955. Amphipoda (Hyperiidea) of the coast of tropical West Africa. *"Atlantide" Report*, 3:7–40.

Repelin, R. 1970. Phronimidae du bassin Indo-Australien (Amphipodes Hypérides). Recherche du cycle génital et examen quantitatif et écologique de la distribution saisonnière. *Cahiers O.R.S.T.O.M., série Océanographique*, II 8(2):65–109.

— 1978. Les Amphipodes pélagiques du Pacifique occidental et central. *Travaux et Documents de l'O.R.S.T.O.M.*, 86, 381 pp.

Richter, G. 1978. Beobachtung zur Entwicklung und Verhalten von *Phronima sedentaria* (Forskål) (Amphipoda Hyperiidea). *Senckenbergiana Maritima*, 10:229–242.

Ruffo, S. 1938. Studi sui Crostacei Anfipodi. VIII: Gli Anfipodi Marini del Museo Civico di Storia Naturale di Genova. (a) Gli Anfipodi del Mediterraneo. *Annali del Museo Civico di Storia Naturale di Genova*, 60:127–151.

Sanger, G. A. 1973. Epipelagic amphipods (Crustacea) off Washington and British Columbia, October–November 1971. *Northwest Fisheries Center. Natural Marine Fishery Service*, *MARMAP* Survey 1, Report 8:1–29.

Schaadt, M. S. 1982. *The origin of the barrel and the food of the mesopelagic hyperiid amphipod crustacean Phronima sedentaria in the Eastern North Pacific Ocean*. M.Sc. Thesis, California State University, Long Beach, California, 55 pp.

Schellenberg, A. 1927. Amphipoda des Nordischen Plankton. *Nordisches Plankton*, 6(20):589–722.

— 1933. Der Brutapparat des pelagischen Amphipoden *Rhabdosoma whitei* Bate. *Zoologischer Anzeiger*, 103:154–158.

Schlieper, C. 1925. Der Farbwechsel von *Hyperia galba*. *Zeitschrift für vergleichende Physiologie*, 3:547–557.

Schmitz, E. H. 1992. Amphipoda. In: F. W. Harrison & A. G. Humes (eds.) *Microscopic anatomy of invertebrates*. Vol. 9: *Crustacea*, New York: Wiley-Liss Inc., pp. 443–528.

Schneppenheim, R. & Weigmann-Haass, R. 1986. Morphological and electrophoretic studies on the genus *Themisto* (Amphipoda: Hyperiidea) from the South and North Atlantic. *Polar Biology*, 6(4):215–225.

Semenova, T. N. 1981. *Parapronöe elongata* sp.n. (Crustacea: Amphipoda Hyperiidea) and discussion of the status of the genus *Sympronöe* Stebbing, 1888. *Zoologicheskii Zhournal*, 60(10):1581–1585. [In Russian]

Senna, A. 1902. Le esplorazioni abissali nel Mediterraneo del piroscafo "Washington" nel 1881. I. Oxycephalidi. *Bollettino della Società Entomologica Italiana*, 34:10–32.

— 1908. Su alcuni Anfipodi Iperini del plancton di Messina. *Bollettino della Società Entomologica Italiana*, 28:153–175.

Sheader, M. 1974. North Sea Hyperiid Amphipoda. *Proceedings of Challenger Society*, 4(5):247.

— 1975. Factors influencing change in the phenotype of the planktonic amphipod *Parathemisto gaudichaudii* (Guérin-Ménéville). *Journal of the Marine Biological Association of the United Kingdom*, 55:132–140.

— 1990. Morphological adaptations permitting resource partitioning in the predatory hyperiid *Themisto* (Amphipoda: Hyperiidea). In: M. Barnes & R. N. Gibson (eds.), *Proceedings of the 24th European Marine Biology Symposium*, Aberdeen, Scotland: University of Aberdeen Press, pp. 478–490.

— & Batten, S. D. 1995. Comparative study of sympatric populations of two hyperiid amphipods, *Primno johnsoni* and *P. evansi*, from the eastern North Atlantic Ocean. *Marine Biology*, 124(1):43–50.

— & Evans, F. 1974. The taxonomic relationship of *Parathemisto gaudichaudii* (Guérin) and *P. gracilipes* (Norman), with a key to the genus *Parathemisto*. *Journal of the Marine Biological Association of the United Kingdom*, 54(4):915–924.

— — 1975. Feeding and gut structure of *Parathemisto gaudichaudii* (Guérin) (Amphipoda: Hyperiidea). *Journal of the Marine Biological Association of the United Kingdom*, 55:641–656.

Shih, Ch.-t. 1969. The systematics and biology of the family Phronimidae (Crustacea: Amphipoda). *"Dana" Reports*, 14(74):1–100.

— 1991. Description of two new species of *Phronima* Latreille, 1802 (Amphipoda: Hyperiidea) with a key to all the species of the genus. *Journal of Crustacean Biology*, 11(2):322–335.

Shoemaker, C. R. 1925. Amphipoda collected by the United States Fisheries steamer "Albatross" in 1911 chiefly in the Gulf of California. *Bulletin of the United States National Museum*, 52:21–61.

— 1930. The Amphipoda of the "Cheticamp" Expedition of 1917. *Contributions to Canadian Biology*, 5:221–359.

— 1945. The Amphipoda of the Bermuda Oceanographic Expeditions, 1929–1931. *Zoologica, Scientific Contribution of the New York Zoological Society*, 30(4):185–266.

Shulenberger, E. 1977. Hyperiid amphipods from the zooplankton community of the North Pacific central gyre. *Marine Biology*, 42:375–385.

Spandl, H. 1924. Amphipoda Hyperiidea aus der Adria. *Zoologischer Anzeiger*, 58:261–272.

— 1924. Expeditionen S. M. Schiff "Pola" in das Rote Meer, die Nördliche und Südliche Hälfte, 1895/96–1897/98, Zoologische Ergebnisse, 35. Die Amphipoden des Roten

Meeres. *Denkschriften der Akademie der Wissenschaften in Wien, Mathematisch-Naturwissenschaftliche Klasse*, 99(Suppl.):19–73.

—1927. Die Hyperiiden (excl. Hyperiidea Gammaroidea und Phronimidae) der Deutschen Südpolar-Expedition 1901–1903. In: *Deutsche Südpolar-Expedition, 1901–1903*, 19 (*Zoologie* 11):145–287.

Spoel, S. van der. 1984. The western distribution border of Phronimidae (Crustacea: Amphipoda: Hyperiidea) in the north Atlantic. *Uttar Pradesh Journal of Zoology*, 4(1):10–16.

Stebbing, T. R. 1888. Report on the Amphipoda collected by H.M.S. "Challenger" during the years 1873–1876. *Report on the Scientific Results. Zoology*, 29(1–3):1–1737.

—1910. General catalogue of South African Crustacea, 5. *Annals of the South African Museum*, 6:281–593.

Stephensen, K. 1918. Hyperiidea–Amphipoda (part 1: Lanceolidae, Scinidae, Vibiliidae, Thaumatopsidae). *Report on the Danish Oceanographical Expeditions, 1908–1910 to the Mediterranean and Adjacent Seas*, 2(D,2):1–70.

—1923. Crustacea, Malacostraca, 5: Amphipoda, 1. *Danish "Ingolf"-Expedition*, 3(8):1–100.

—1924. Hyperidea–Amphipoda (part 2: Paraphronimidae, Hyperiidae, Dairellidae, Phronimidae, Anchylomeridae). *Report on the Danish Oceanographical Expeditions, 1908–1910, to the Mediterranean and Adjacent Seas*, 2(D,4):71–149.

—1925. Hyperiidae–Amphipoda (part 3: Lycaeopsidae, Pronoidae, Lycaeidae, Brachyscelidae, Oxycephalidae, Parascelidae, Platyscelidae). *Report on the Danish Oceanographical Expeditions, 1908–1910 to the Mediterranean and Adjacent Seas*, 3(D,5):151–252.

—1928. Contribution à l'étude de la faune du Cameroun. Crustacea. III. Amphipoda. *Faune des Colonies Françaises*, Paris, 1(6):589–591.

—1944. The Zoology of East Greenland. Amphipoda. *Meddelelser om Grønland*, 121(14):1–165.

—1947. Tanaidacea, Isopoda, Amphipoda, and Pycnogonida. In: *Scientific Results of the Norwegian Antarctic Expeditions, 1927–1928*, 27:1–90.

—& Pirlot, J.-M. 1931. Les Amphipodes Hypérides du genre *Mimonectes* Bovallius (includ. *Sphaeromimonectes* Woltereck et *Parascina* Stebbing) et de quelques genres voisins (*Archaeoscina* Stebbing, *Micromimonectes* Woltereck, *Microphasma* Woltereck et *Proscina* n.g.). *Archives de Zoologie Expérimentale et Générale*, 71(4):501–553.

Steuer, A. 1911. Adriatische Planktonamphipoden. *Sitzungsberichte der Kaiserlichen Akademie der Wissenschaften in Wien, Mathematisch- Naturwissenschaftliche Klasse*, 120(6):671–688.

Stewart, D. A. 1913. A report on the extra-antarctic Hyperiidea collected by the "Discovery". *Annals and Magazine of Natural History, Series 8*, 12:245–265.

Stout, V. 1913. Studies in Laguna Amphipoda. II. *Zoologische Jahrbücher. Abteilung für Systematik, Ökologie und Geographie der Tiere*, 34:633–659.

Streets, T. 1877. Pelagic Amphipoda. *Proceedings of the Academy of Natural Sciences of Philadelphia*, pp. 276–290.

—1877. Contributions to the natural history of the Hawaiian and Fanning Islands and Lower California. *Bulletin of the United States Natural History Museum*, 7:1–172.

Tashiro, J. E. & Jack, W. J. 1972. Amphipoda (Hyperidea) Distribution and abundance off the coast of Central West Africa. *NOAA, National Marine Fisheries Service, Seattle, Washington, Data Report*, 76, 38 pp.

Tattersall, W. M. 1906. The marine fauna of the coast of Ireland. 8. Pelagic Amphipoda of the Irish Atlantic slope. *Fisheries Branch, Ireland Scientific Investigations, 1905.* No. 4:1–39.

Thorsteinsson, E. D. 1941. New or noteworthy amphipods from the North Pacific Coast. *University of Washington Publications in Oceanography*, 4(2):50–94.

Thurston, M. H. 1973. On the identity of *Lanceola aestiva* Stebbing, 1888. (Amphipoda, Lanceolidae). *Crustaceana*, 24(3):334–336.

—1976. New pelagic amphipods collected on the SOND cruise. *Journal of the Marine Biological Association of the United Kingdom*, 56(1):143–159.

—1976. The vertical distribution and diurnal migration of the Crustacea Amphipoda collected during the SOND cruise, 1965. II. The Hyperiidea and general discussion. *Journal of the Marine Biological Association of the United Kingdom*, 56(2):383–470.

—1977. Depth distribution of *Hyperia spinigera* Bovallius, 1889 (Crustacea, Amphipoda, Hyperiidea) and medusae in the North Atlantic Ocean, with notes on the associations between *Hyperia* and coelenterates. In: M. Angel (ed.), *A voyage of discovery — George Deacon 70th anniversary volume*, Oxford: Pergamon Press Ltd., pp. 499–536.

Trégouboff, G. & Rose, M. 1957. *Manuel de Planctonologie Méditerranéenne*, Vol. 1, 587 pp.; Vol. 2, 207 Plates. Paris: Centre National de la Recherche Scientifique.

Vinogradov, G. M. 1988. Life forms of amphipods hyperiids *Hyperia* and *Parathemisto* at different stages of ontogeny. *Zoologicheskii Zhurnal*, 67(3):346–352. [In Russian]

—1990. Rare and new hyperiid species for the Indian Ocean. *Trudy Instituta Okeanologii Akademii Nauk SSSR*, 124:105–111. [In Russian]

—1990. Amphipods in pelagial south-eastern area of Pacific Ocean. *Trudy Instituta Okeanologii Akademii Nauk SSSR*, 124:27. [In Russian]

—1990. Life form ratio of hyperiid amphipods in different part of the ocean. *Okeanologiya*, 30(4):656–665. [In Russian]

—1991. On *Tetrathyrus arafurae* (Amphipoda: Hyperiidea: Platyscelidae) found in the Indian Ocean. *Vestnik Zoologii*, 5:81. [In Russian]

Vinogradov, M. E. 1956. Hyperiids (Amphipoda: Hyperiidea) of the western Bering Sea. *Zoologicheskii Zhurnal*, 35(2):194–218. [In Russian]

—1957. Hyperiids (Amphipoda) of the Northwest Pacific Ocean. *Trudy Instituta Okeanologii Akademii Nauk SSSR*, 20:186–227. [In Russian]

—1960. Hyperiidea Physosomata of the tropical Pacific Ocean. *Trudy Instituta Okeanologii Akademii Nauk SSSR*, 41:198–247. [In Russian]

—1962. Hyperiids (Amphipoda, Hyperiidea) collected by the Soviet Antarctic Expedition on M/V "Ob" South of 40°S. *Studies of marine fauna I (IX)*. In: *Biological results of the Soviet Antarctic Expedition (1955–1958)*, 1:5–35. Zoological Institute, Academy of Sciences SSSR. [In Russian] (English translation by Israel Program for scientific translations published in 1966).

—& Semenova, T. N. 1985. A New species of *Hyperia* (Crustacea: Amphipoda: Hyperiidea) from coastal waters of Peru. *Zoologicheskii Zhurnal*, 64(1):139–143. [In Russian]

—, Volkov, A. F. & Semenova, T. N. 1982. *Amphipods – Hyperiids (Amphipoda, Hyperiidea) of the World Ocean*. Leningrad: "Nauka", 491 pp. [In Russian]

Vives, F. 1966. Zooplancton nerítico de las aguas de Castellón (Mediterráneo Occidental). *Investigacion Pesquera*, 30:49–166.

— 1968. Sur les Malacostracés planctoniques des mers Tyrrhénienne et Catalane. In: *Rapports et Procès-Verbaux des Réunions, Comission Internationale pour l'Exploration Scientifique de la Mer Méditerranée*, 19(3):459–461.

—, Santamaría, G. & Trepat, I. 1975. El zooplancton de los alrededores del Estrecho de Gibraltar en junio–julio de 1972. *Resultados Expediciones Cientificas del Buque Oceanografico "Cornide de Saavedra"*, 4:7–100.

Vosseler, I. 1900. Die verwandtschaftliche Beziehungen der Sciniden und eine neue Gattung derselben (*Acanthoscina*). *Zoologischer Anzeiger*, 23:662–676.

— 1901. Die Amphipoden der Plankton-Expedition, I. Hyperiidea, 1. *Ergebnisse der Plankton-Expedition der Humboldt-Stiftung*, 2(G,e):1–129.

Wagler, E. 1926. Amphipoda. 2: Scinidae. *Ergebnisse der Deutschen Tiefsee-Expedition auf dem Dampfer "Valdivia", 1898–1899*, 20(6):317–446.

— 1927. Die Sciniden der Deutschen Südpolar-Expedition, 1901–1903. *Deutsche Südpolar-Expedition, 1901–1903*, 19:85–111.

Walker, A. O. 1903. Amphipoda of the "Southern Cross" Antarctic Expedition. *Journal of the Linnean Society of London*, 29:38–64.

— 1907. Crustacea. III: Amphipoda. In: *Natural Antarctic Expedition, 1903–1904*, British Museum (Natural History) 3:1–39.

Weigmann-Haass, R. 1989. Zur Taxonomie und Verbreitung der Gattung *Hyperiella* Bovallius 1887, im antarktischen Teil des Atlantik (Crustacea: Amphipoda: Hyperiidea). *Senckenbergiana Biologica*, 69(1–3):177–191.

— 1989. Taxonomie und Verbreitung von *Vibilia antarctica* Stebbing 1888 im antarktischen Teil des Atlantik (Crustacea: Amphipoda: Hyperiidea). *Senckenbergiana Biologica*, 70 (4–6):419–428.

— 1991. Zur Taxonomie und Verbreitung der Gattung *Hyperoche* Bovallius 1887 im antarktischen Teil des Atlantik (Crustacea: Amphipoda: Hyperiidea). *Senckenbergiana Biologica*, 71(1–3):169–179.

Westernhagen, von, H. 1976. Some aspects of the biology of the hyperiid amphipod *Hyperoche medusarum*. *Helgoländer wissenschaftliche Meeresuntersuchungen*, 28:43–50.

White, M. G. & Bone, D. C. 1972. The interrelationship of *Hyperia galba* (Crustacea, Amphipoda) and *Desmonema gaudichaudii* (Scyphomedusae, Semaeostomae) from the Antarctic. *Britain Antarctic Survey Bulletin*. 27:39–49.

Willemöes-Suhm, R. 1875. On some Atlantic Crustacea from the "Challenger" Expedition. *Transactions of the Linnean Society of London, Series 2, Zoology*, 11(1):23–59.

Wing, B. L. 1976. *Ecology of Parathemisto libellula and P. pacifica (Amphipoda: Hyperiidea) in Alaskan coastal waters*. Ph.D. Thesis. University of Rhode Island. Kingston.

Woltereck, R. 1903. Bemerkungen zu den Amphipoda Hyperiidea der Deutschen Tiefsee-Expedition. I. Thaumatopsidae. *Zoologisher Anzeiger*, 26(7):447–459.

— 1904. Zweite Mitteilung über die Hyperiden der Deutschen Tiefsee-Expedition.

"Physosoma", ein neuer pelagischer Larventypus, nebst Bemerkungen zur Biologie von *Thaumatops* und *Phronima. Zoologischer Anzeiger,* 27(18):553–563.

—1904. Dritte Mitteilung über Hyperiden der Deutschen Tiefsee-Expedition und erste Notiz über die Amphipoden der Deutschen Südpolar-Expedition. *Zoologischer Anzeiger,* 27(20–21):621–629.

—1905. Mitteilungen über Hyperiden der Deutschen Tiefsee-Expedition und erste Notiz über die Amphipoden der Deutschen Südpolar-Expedition. *Scypholanceola,* eine neue Hyperidengattung mit Reflectororganen. *Zoologischer Anzeiger,* 29(13):413–417.

—1906. Fünfte Mitteilung über die Hyperiden der "Valdivia" Expedition. *Mimonectes* n. gen., *Zoologischer Anzeiger,* 30(7):187–194.

—1906. Weitere Mitteilung über Hyperiden der "Valdivia" (N6) und "Gauss"-Expedition (N3): *Sphaeromimonectes scinoides* (n. sp.), *S. gaussi, S. cultricornis* (n. sp.) *und S. valdiviae,* eine morphologische Reihe. *Zoologischer Anzeiger,* 30(26):865–869.

—1907. Siebente Mitteilung über die "Valdivia"-Hyperiden: *Prolanceola vibiliformis* nov. gen. nov. sp., *Zoologischer Anzeiger,* 31(5/6):129–132.

—1909. Amphipoda. Reports on the scientific results of the expedition to the Eastern Tropical Pacific. United States Fishery Commission Steamer "Albatross", from October 1904 to March 1905. *Bulletin of the Museum of Comparative Zoology, Harvard,* 52(9):145–168.

—1927. Die Lanceoliden und Mimonectiden der Deutschen Südpolar-Expedition 1901–1903. 19 (*Zoologie,* 11):57–84.

Yang, W. T. 1960. A study of the subgenus *Parahyperia* from the Florida Current (genus *Hyperia,* Amphipoda: Hyperiidae). *Bulletin of Marine Sciences of the Gulf and Caribbean,* 10(1):11–39.

Yoo, K. I. 1971. Pelagic Hyperiids (Amphipoda–Hyperiidea) of the Western North Pacific Ocean. *Journal of the National Academy of Sciences, Republic of Korea, Natural Sciences Series,* 10:39–89.

—1971. The biology of the pelagic amphipod, *Primno macropa* Guér. in the Western North Pacific. 1. Systematics. *Korean Journal of Zoology,* 14(3):132–138.

—1972. The biology of the pelagic amphipod *Primno macropa* Guér. in the Western North Pacific. 2. Geographical distribution and vertical distributional pattern. *Korean Journal of Zoology,* 15(2):87–91.

Young, J. W. 1989. The distribution of hyperiid amphipods (Crustacea: Peracarida) in relation to warm-core eddy J in the Tasman Sea. *Journal of Plankton Research,* 11:711–728.

—& Anderson, D.T. 1987. Hyperiid amphipods (Crustacea: Peracarida) from a warm-core eddy in the Tasman Sea. *Australian Journal of Marine and Freshwater Research,* 38:711–725.

Zeidler, W. 1978. Hyperiidea (Crustacea: Amphipoda) from Queensland waters. *Australian Journal of Zoology, supplemental series,* 59:1–93.

—1984. Distribution and abundance of some Hyperiidea (Crustacea: Amphipoda) in Northern Queensland waters. *Australian Journal of Marine and Freshwater Research,* 35:285–305.

—1990. Pelagic Amphipods. Infraorder Physosomata (Crustacea: Amphipoda: Hyperiidea) from the CSK International Zooplankton Collection (Western North Pacific), with the description of four new species of *Scina*. *Publications of the Seto Marine Biology Laboratory*, 34(4/6):167–200.

— 1991. Crustacea Amphipoda Hyperiidea from MUSORSTOM cruises. In: A. Crosnier (ed.), *Résultats des campagnes MUSORSTOM, 9. Mémoires du Muséum National d'Histoire Naturelle, sér. A*, 152:125–137.

— 1992. Hyperiid Amphipods (Crustacea: Amphipoda: Hyperiidea) collected recently from Eastern Australian waters. *Records of Australian Museum*, 44:85–133.

— 1992. A new species of pelagic amphipod of the genus *Lestrigonus* (Crustacea: Amphipoda: Hyperiidea: Hyperiidae) from eastern Australia. *Journal of Plankton Research*, 14(10):1383–1396.

Zelickman, E. & Por, F.D. 1996. Ultrastructure of the pereiopodal dactyls in the family Phronimidae Dana, 1852 (Crustacea: Amphipoda: Hyperiidea). *Journal of Natural History, London*, 30(8):1193–1213.

כתבי האקדמיה הלאומית הישראלית למדעים

החטיבה למדעי-הטבע

החי של ארץ-ישראל

סרטנים 1: אטלס של היפרידיאה (Hyperiidea) מן הים התיכון ומים סוף

מאת

אנגלינה א' זליקמן

ירושלים תשס"ה